ADVANCES IN UNDERWATER ACOUSTICS

Edited by **Andrzej Zak**

Advances in Underwater Acoustics
http://dx.doi.org/10.5772/65227
Edited by Andrzej Zak

Contributors

Sara Pensieri, Roberto Bozzano, Vijaya Baskar V, Henry M. Manik, S.M. Jiang, Qian Lu, Liu Feng, Ying Zhang, Robert N.K. Loh, Salah Bourennane, Caroline Fossati

Published by InTech

Janeza Trdine 9, 51000 Rijeka, Croatia

Publishing Process Manager Iva Simcic
Technical Editor SPi Global
Cover InTech Design team

First published November, 2017
Printed in Croatia
Legal deposit, Croatia: National and University Library in Zagreb

Additional hard copies can be obtained from orders@intechopen.com

Advances in Underwater Acoustics , Edited by Andrzej Zak
p. cm.
Print ISBN 978-953-51-3609-5
Online ISBN 978-953-51-3610-1

For EU product safety concerns:
IN TECH d.o.o., Prolaz Marije Krucifikse Kozulić 3, 51000 Rijeka, Croatia,
info@intechopen.com or visit our website at intechopen.com.

Meet the editor

Prof. Andrzej Zak comes from Poland. He graduated from the Military University of Technology with a Master of Science degree in Engineering in the field of Computer Science. In the Polish Naval Academy, Prof. Zak did his degree of Doctor of Technical Sciences. AGH University of Science and Technology awarded him degree of doctor habilitated in technical science in the field of automation and robotics. He started his professional career in the Hydroacoustics Department of the Polish Naval Academy where he dealt with hydroacoustic measurements and methods of signals processing including applying artificial intelligence to the recognition of sailing objects. At present, his interests are focused on wireless data transmission in water. Prof. Zak has published many articles and conference papers covering topics related directly or indirectly to underwater acoustics.

Contents

Preface

From the earliest times, humanity, pushed by curiosity, seeks ways to observe the depths of the oceans, seas, and inland waters. Our senses, adapted to the physical properties of the environment in which we live, that is, the surface of the earth and the surrounding atmosphere, generally do not work when we want to use them underwater. In addition, the human body is not adapted to unfavorable underwater conditions, especially at deep depths. Therefore, the only possibility is to use technical means of observation that will not only replace our senses but will also allow us to increase the area of observation and provide data that are not available in the environment to which we are adapted. To obtain information or to transmit data on the surface of the earth and in the surrounding atmosphere, we most often use electromagnetic waves, primarily because of their low attenuation. Unfortunately, the damping of electromagnetic waves in water is very strong, and therefore they cannot be used in this environment. Fortunately, the acoustic waves are much less attenuated, so they perform the same functions as electromagnetic waves in the air. Acoustic waves have several important features that make them a basic underwater observation tool and information medium. These include, among others, low attenuation of low-frequency acoustic waves; ease of radiation and reception of narrow beam of waves; and propagation over long distances especially at low frequency, due to the difference in propagation velocity of the acoustic wave in water compared to the electromagnetic wave in the air, less frequency is required to achieve a similar wavelength.

Underwater acoustics is a part of acoustics dealing with the propagation of elastic waves in the water and their production and reception in the frequency range of sound and ultrasound. It should be emphasized that the range of interest covers not only technical aspects but also biological, medical, and much more. Hence, dealing with underwater acoustics requires appropriate knowledge and skills most often resulting from several areas of science and technology. Underwater acoustics, because of its interdisciplinary character, is one of the most productive fields of acoustics and one that stimulates useful interchanges between different disciplines of science and attracts the attention of a large number of researchers all over the world.

The beginnings of underwater acoustics are related to the development of shipping, which occurred in the second half of the nineteenth century, after the introduction of steam propulsion of ships. More frequent collisions with navigational obstacles have resulted in increasing financial and, above all, human losses. It is commonly believed that the direct contribution to the construction of the first underwater acoustic device was the Titanic disaster. As a result of this incident in 1917, Pierre Langevin built the first sonar to detect icebergs and sea bottom observation. In his solution, he used the piezoelectric effect discovered by the Curie brothers in 1880. This allowed for the efficient conversion of electrical signals to acoustic wave and vice versa. Over time, using increasingly improved materials, the natural barrier of conversion of electric to acoustic energy has been minimized. Solving the problem of energy conversion with rapid advances in electronic technology, especially digital signal analysis, has enabled the development of efficient and compact underwater acoustic devices. The main motive behind the development of hydroacoustic systems was, as usual, military needs. The first qualitative progress was made in World War II and was directly related to the efficient and massive use of submarines by Germany in the battle for supremacy in the seas and oceans. Another significant progress was made during the Cold War and was directly related to the threat posed by submarines equipped with missiles with nuclear warheads. At present, this is due to the threat of terrorism and the related need to detect small submarines, scuba divers, mines, and other objects threatening marine transport and infrastructure. Not without significance is the development of digital technology that allows the use of a wider range of computational tools, thus achieving a new quality in processing of acquired data. More and more important is the need for underwater communication, especially in regions that are detrimental to the propagation of sound waves, such as shallow waters, channels, harbors, etc.

Although advances in underwater acoustics are primarily based on military needs, the results are rapidly transferring to the civilian sphere of life. Hydroacoustic systems are now widely used in navigation, fishing, ocean engineering, hydrography, or biology. According to the maritime law, each vessel must have a navigation echo sounder on its equipment. The group directly interested in the use of underwater acoustic systems is very large. Besides the ship's commanders, merchant fleet officers, fishing boat shippers, or divers, the scientists use broadly underwater acoustic systems in marine researches. It is also used by the institutions involved with the maintenance of the fairways, harbors, channels, as well as government and nongovernmental institutions dealing with internal and external security. This collection is complemented by police or firefighters, as well as employees of underwater engineering companies that extract seabed raw materials and employees of commercial and advisory companies operating in the submarine acoustic industry.

The main objective of this book is to present the latest achievements in the field of underwater acoustics and a better understanding of the involved key issues. Moreover, the presented practical solutions of problems allow the same to become familiar with the technical and nontechnical aspects of various applications as well as the latest developments in underwater acoustic systems used for exploration of the marine environment. For this reason, scientists from research institutions and universities of various countries of the world present their achievements in the field of underwater acoustics. According to the information presented in this book, the impact of these systems on the improvement of human life is invaluable in particular due to the wide field of application. I expect that this book will bring some of the problems associated with underwater acoustics and will be an inspiration for further research. I also hope that this publication will be received by the acoustic community with great interest.

Andrzej Zak
Polish Naval Academy
Poland

Chapter 1

Localization of Buried Objects Using Reflected Wide-Band Underwater Acoustic Signals

Salah Bourennane and Caroline Fossati

Additional information is available at the end of the chapter

http://dx.doi.org/10.5772/intechopen.71272

Abstract

This chapter deals with the localization of wide-band underwater acoustic sources. A combination of high resolution methods with scattering acoustic model are proposed. The bearing and the range sources at each sensor are expressed as a function to those at the first sensor. We present the noneigendecomposition methods fixed-point algorithm, projection approximation subspace tracking (PAST) algorithm, PAST with deflation (PASTD) algorithm and orthogonal PAST (OPAST) algorithm to track the signal subspace to compute leading eigenvectors. The proposed algorithms are faster than singular value decomposition (SVD) for MUSIC. The spatial smoothing operator is used to decorrelate the received signals and to estimate the coherent signal subspace. The performance of the different methods are evaluated by both computer simulations and experimental and data recorded during underwater acoustic experiments.

Keywords: array processing, source localization, wide-band, fast algorithm

1. Introduction

Non-invasive detection and localization of sources is an important application area in many application domains, such as radar, sonar, seismology and communications. Thus there has been a growing interest in developing techniques for the estimation wavefronts of the direction-of-arrival (DOA) in order to detect and localize the emitting sources [1]. Support vector machine (SVM) based on electromagnetic approach [2–4] and conventional neural networks (NN) based on inverse scattering technique [5] are proposed for buried object detection. Ground penetrating radar (GPR) is used to improve the detection of weak-scattering plastic mines [6]. But electromagnetic filed inversion require more computational effort. The inversion of measured scattered acoustical waves is used to image buried objects, but it needs high frequencies and the application in a real environment is difficult [7]. Therefore, the acoustic imagery technique is not suitable because the high frequencies are strongly attenuated inside the sediment. Using a low frequency, synthetic aperture sonar (SAS) has been recently applied on partially and shallowly buried cylinders in a sandy seabed [8]. The bearing and the range estimation using correlated signals scattered from nearfield and farfield objects, in a noise environment, still a challenging problem. The MUSIC algorithm is one of the most thoroughly studied and best understood subspace based high resolution methods. It divides the observation space into two signal-subspaces: the signal subspace and the noise subspace [9]. MUSIC uses the orthogonality property between the two areas to locate sources. Different approaches exist to detect and localize buried objects but acoustic techniques will be considered in our study. Match field processing (MFP) [10] has been successfully used for localization sources in ocean acoustic. We discuss the proposed approach based on MUSIC associated with acoustic scattering model referred to MFP [10]. We take into account the water-sediment interface [11]. This means that we attempt to combine both the reflection and refraction of wave in the model [12]. From the exact solution of the acoustic scattered field [13], we have derived a new source steering vector including both the ranges and the bearings of the objects. This source steering vector is employed in objective function instead of the classical plane wave model [14, 15] which have extended the 1-D MUSIC to 2-D MUSIC. The acoustic scatter field model has been addressed in many published researches with different configurations. For example, the configurations can be single [16] or multiple objects [17], buried or partially buried objects [18] with cylindrical [16] or spherical shape [19]. All those scattering models can be used with the proposed source steering vector. In this chapter a spatial smoothing operator is proposed to estimate the coherent signal subspace [20]. Inverse power method, which allows to find an approximate eigenvector when an approximation to corresponding eigenvalue is already known, is proposed to estimate the required noise variance. In high resolution method, we use singular value decomposition (SVD) in music for obtaining the eigenvectors noise subspace. However, the main drawback is the inherent complexity and computational time load [21]. So a large number of approaches have been introduced for fast subspace tracking in order to overcome this difficulty. We propose to replace SVD by Fixed Point for computing leading eigenvectors from the spectral matrix [22, 23]. We propose another methods to accelerate MUSIC, such as projection approximation subspace tracking (PAST) [24, 25], which makes the expectation of square difference between the input vector and the projected vector minimum. With proper projection approximation, the PAST derives a recursive least squares (RLS) algorithm for tracking the signal subspace. The PAST algorithm computes an asymptotically orthogonal basis of the signal subspace. PAST with deflation (PASTD) is derived from PAST by applying the deflation technique in order to get the signal eigenvectors and eigenvalues [24, 26]. It has been shown that these subspace trackers are closely linked to the classical power iterations method, but does not guarantee the orthonormality at each iteration [27, 28]. Orthogonal PAST (OPAST) algorithm is another fast implementation of the power method which outperforms both PAST and PASTD to reduce computation time [29, 30]. The performance of the proposed algorithms are evaluated by several numerical simulations and the data has been recorded using an experimental water tank.

The remainder of the chapter is as follows: Section 2 introduces the problem formulation. Section 3 presents the scattering acoustic model of generating the received signals. Then proposed algorithm for fast localization of underwater acoustic in presence of correlated noise is presented in Section 4. Section 5 proposes the new versions of "MUSIC" without eigendecomposition. Some numerical results and Experimental tests are shown in Sections 6 and 7, respectively. Finally, Section 8 summarizes the main conclusions of this chapter.

Throughout the chapter, we use to denote: transpose operation "T," complex conjugate transpose "+, " complex conjugate "∗," expectation operator $E[.]$, cumulant $Cum(.)$, Kronecker product ⊗, determinant $|\cdot|$ and Frobenius norm $\|\cdot\|_F$.

2. Problem formulation

Consider a transmitter that generates a plane wave with an angle θ_{inc}. The incident plane wave will propagate and be reflected by P objects. For example, when it is located in the bottom of a tank filled with sand and water. We name the objects, which reflect the signals, is the sources. An array composed of N sensors receives K signals emitted by the sources ($P < N$). The received signals are grouped into a vector $\mathbf{r}(f)$, which is the Fourier transform of the array output vector at frequency f, is written as [31, 32]:

$$\mathbf{r}(f) = \mathbf{A}(f)\mathbf{s}(f) + \mathbf{b}(f) \tag{1}$$

where $\mathbf{A}(f) = [\mathbf{a}(f, \theta_1), \mathbf{a}(f, \theta_2), \dots, \mathbf{a}(f, \theta_K)]$, matrix of dimensions $(N \times P)$ is the transfer matrix of the source-sensor array systems with respect to some chosen reference point, $\mathbf{s}(f) = [s_1(f), s_2(f), \dots, s_K(f)]^T$ is the vector of signals, $\mathbf{b}(f) = [b_1(f), b_2(f), \dots, b_K(f)]^T$ is the vector of Gaussian white noise. We define the matrix interspersals by:

$$\mathbf{\Gamma}(f) = E[\mathbf{r}(f)\mathbf{r}^+(f)] \tag{2}$$

This matrix is estimated by $\widehat{\mathbf{\Gamma}}(f) = \frac{1}{L_r}\sum_{l=1}^{L_r} \mathbf{r}_l(f)\mathbf{r}_l^+(f)$ where L_r represents the number of realizations. Thus the spectral matrix $\mathbf{\Gamma}(f)$ is formed:

$$\mathbf{\Gamma}(f) = \mathbf{A}(f)\mathbf{\Gamma}_s(f)\mathbf{A}^+(f) + \mathbf{\Gamma}_b(f), \tag{3}$$

where $\mathbf{\Gamma}_b(f) = E[\boldsymbol{b}(f)\boldsymbol{b}^+(f)]$ is the spectral matrix of noise vector, the spectral matrix of signal vector is given as:

$$\mathbf{\Gamma}_s(f) = E[\mathbf{s}(f)\mathbf{s}^+(f)] = \mathbf{V}(f)\mathbf{\Lambda}(f)\mathbf{V}^+(f) + \mathbf{\Gamma}_b(f) \tag{4}$$

where $\mathbf{\Lambda}(f) = diag\{\lambda_1(f), \dots, \lambda_P(f)\}$ and $\mathbf{V}(f) = [\mathbf{v}_1(f), \dots, \mathbf{v}_P(f)]$. Assuming that the columns of $\mathbf{A}(f)$ are linearly independent, in other words, $\mathbf{A}(f)$ is full rank, it follows that for nonsingular $\mathbf{\Gamma}_s(f)$, the rank of $\mathbf{A}(f)\mathbf{\Gamma}_s(f)\mathbf{A}^+(f)$ is P. This rank property implies that:

- the $(N - P)$ multiplicity of its smallest eigenvalues: $\lambda_{P+1}(f) = \dots = \lambda_N(f) \cong \sigma^2(f)$.
- the eigenvectors corresponding to the minimal eigenvalues are orthogonal to the columns of $\mathbf{A}(f)$, namely, $\mathbf{V}_b(f)$ is equal to by the definition of $\{\mathbf{V}_{P+1}(f)\dots\mathbf{V}_N(f)\}$ orthogonal to $\{\mathbf{a}(f, \theta_1)\dots\mathbf{a}(f, \theta_P)\}$.

The eigenstructure-based techniques are based on the exploitation of these properties. When the objects are far away from the array, the wavefront is assumed to be plane. Then DOA of the sources are obtained, at the frequency f, by the peak positions in a so-called spectrum (MUSIC) defined as:

$$\mathbf{Z}(f,\theta) = \frac{1}{\mathbf{a}^{+}(f,\theta)\mathbf{V}_b(f)\mathbf{V}_b^{+}(f)\mathbf{a}(f,\theta)} \tag{5}$$

where $\mathbf{a}(f,\theta) = \left[1, e^{-2j\pi f \frac{d\sin(\theta)}{c}}, \ldots, e^{-2j\pi f(N-1)\frac{d\sin(\theta)}{c}}\right]$ is the steering vector of plane wave model, $\mathbf{V}_b$ is the eigenvectors of the noise subspace, c is the sound speed, d is the interspacing of the sensors and j is the complex operator. In the presence of P objects, the 1-D $MUSIC(f,\theta)$ algorithm cannot solve all the P angles because the signals are correlated. In the following sections, we use simultaneously all the information contained in the signals to estimate the coherent signal subspace which extends the conventional 1-D $MUSIC(f,\theta)$ algorithm to 2-D $MUSIC(f,\theta,\rho)$ for joint range ρ and DOA θ estimation when the objects are buried in the sand with small depth.

3. Scattering acoustic model: to generate the received signals

In this section, we will present how to fill the vector of the scattering model. We consider a sedimentary covered with water and the interface is treated as a plane. An object of cylindrical or spherical shell is buried in the sediment. An incident plane wave propagating in the water reaches the interface with an angle of incidence θ_{inc} as show in **Figure 1**. The incident plane wave generates a wave reflecting plane in the water and refracted plane wave propagating in the sediment. So the array located in the water receives three components [18]:

- the incident plane wave,
- the reflecting plane wave,
- the transmitter plane wave diffused by the object.

The array-interface height h and the nature of the sediment are known or can be determined. Also, the speed of wave propagation in the sediment c_2 is assumed to be known. Because the object is buried, the pressure in the water and sediment will not be expressed directly in terms of θ_1 and ρ_1, but in terms of five unknown parameters θ_{11}, ρ_{11}, θ_{12}, ρ_{12} and y_c (the depth of buried object). So we will express θ_{11}, ρ_{11}, θ_{12}, ρ_{12} and y_c based on θ_1 and ρ_1 (see **Figure 2**). We use the law of Snell-Descartes and generalize the Pythagorean theorem to obtain the expressions:

$$y_c = \rho_1 \cos(\theta_1) - h \tag{6}$$

$$\theta_{12} = \arcsin\left(\frac{c_2}{c_1}\sin(\theta_{inc})\right) \tag{7}$$

http://dx.doi.org/10.5772/intechopen.71272

$$\rho_{12} = \frac{\rho_1 \cos(\theta_1) - h}{\cos\left[\arcsin\left(\frac{c_2}{c_1}\sin(\theta_{inc})\right)\right]} \tag{8}$$

$$\theta_{11} = \arctan\left(\frac{\rho_1 \cos(\theta_1) - \rho_{12}\cos(\theta_{12})}{\rho_1 \sin(\theta_1) - \rho_{12}\sin(\theta_{12})}\right) \tag{9}$$

$$\rho_{11} = \frac{h}{\cos\left[\arctan\left(\frac{\rho_1 \cos(\theta_1) - \rho_{12}\cos(\theta_{12})}{\rho_1 \sin(\theta_1) - \rho_{12}\sin(\theta_{12})}\right)\right]} \tag{10}$$

3.1. Cylindrical shell

Assume a cylindrical shell long enough which is buried in the sediment with axis parallel to the interface plane. Thus the acoustic pressure wave received by the first sensor of the array $P_{cyl}(f, \theta_{k1}, \rho_{k1})$ contains three acoustic pressure components:

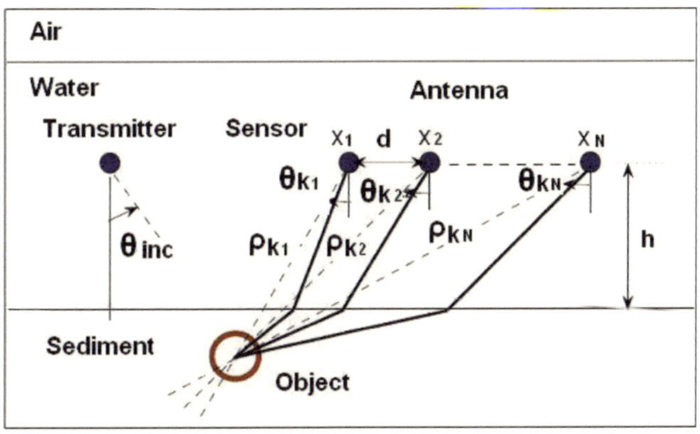

Figure 1. Geometry configuration of buried object.

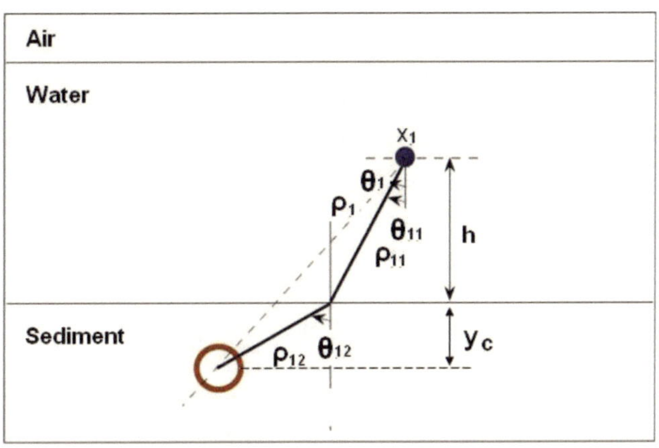

Figure 2. Configuration of the buried object-1st sensor.

$$P_{cyl}(f,\theta_{k1},\rho_{k1}) = P_{inwater-cyl} + P_{refwater-cyl} + P_{diff-cyl} \quad (11)$$

where $P_{inwater-cyl} = e^{jk_1(-(\rho_{k1}\sin(\theta_{k1}))\sin(\theta_{inc}) + h\cos(\theta_{inc}))}$ is the pressure incident in the water, $P_{refwater-cyl} = R(\theta_{inc})e^{jk_1((\rho_{k1}\sin(\theta_{k1}))\sin(\theta_{inc}) - h\cos(\theta_{inc}))}$ is the pressure reflected by the sediment-water interface, where $R(\theta_{inc})$ is the reflection coefficient of the interface, $P_{diff-cyl} = \sum_{m=-\infty}^{+\infty} \xi \mathbf{T}_c(\mathbf{I} - \mathbf{D}_c)^{-1}\Psi^t_{cyl}$ is the diffused acoustic pressure wave transmitted in the water, where $\mathbf{I}$ is the identity matrix, $\mathbf{D}_c$ is a linear operator, $\mathbf{T}_c$ is the transition diagonal matrix, Ψ^t_{cyl} is the vector of transmitted wave and $\xi = [\xi_1, \xi_2, \ldots, \xi_m]$ is defined by $\xi_m = T_{water-sed}(\theta_{inc})e^{jk_2 y_c \cos(\theta_{k11})} j^m e^{-jm(\pi - \theta_{k11})}$, where $T_{water-sed}(\theta_{inc})$ is the transmission coefficient.

3.2. Spherical shell

In this section $P_{sph}(f, \theta_{k1}, \rho_{k1})$ is the acoustic pressure wave received by the first sensor and is expressed as follows [12, 18]:

$$P_{sph}(f,\theta_{k1},\rho_{k1}) = P_{inwater-sph} + P_{refwater-sph} + P_{diff-sph} \quad (12)$$

where $P_{inwater-sph} = \sum_{m=-\infty}^{+\infty} j^m \left[P_m(\cos(\theta_{inc}))e^{2j\cos(\theta_{inc})y_c}\right]$ is the incident wave generates in the water, $P_{refwater-sph} = \sum_{m=-\infty}^{+\infty} R(\theta_{inc})P_m(-\cos(\theta_{inc}))e^{2j\cos(\theta_{inc})y_c}$ is the reflected wave, where $R(\theta_{inc})$ is the reflection coefficient of the interface. We define the acoustic wave of the diffused by spherical shell by $P_{sphsediment} = \left(\mathbf{T}_s^{-1} - \mathbf{C}^m\right)^{-1} P_{inwater-sph}$, where $\mathbf{T}_s$ is the transition matrix and $\mathbf{C}$ is the matrix containing the conversion coefficients. The diffused acoustic pressure wave transmitted in the water is given by: $P_{diff-sph} = \sum_{m=0}^{\infty} \varepsilon_m Y \cos(m(\theta_{k1} - \theta_{12}))$, where $Y = [P_{sphsediment} P_m(\cos(\theta_{k1}))h_m(k_2\rho_{k1}) + X_m]$, $\varepsilon_m = 2$ for $m > 0$, $\varepsilon_0 = 1$.

The vector $\mathbf{a}(f, \theta_K, \rho_K)$ is filled with cylindrical or spherical scattering model considering the sources shape. For example, when the sources are cylindrical shells, the vector is given by:

$$\mathbf{a}(f,\theta_K,\rho_K) = \left[P_{cyl}(f,\theta_{k1},\rho k1),\ldots,P_{cyl}(f,\theta_{kN},\rho_{kN})\right] \quad (13)$$

Eq. (11) or (12) give the first component of the vector. The other $P_{cyl}(f, \theta_{ki}, \rho ki)$ for $i = 1, 2, \ldots, N$ associated with the ith sensors can be formed by a geometric recursive relationship. The relationship allows to express (θ_{ki}, ρ_{ki}) according to $(\theta_{ki-1}, \rho_{ki-1})$ (see **Figure 1**). This recursive calculate is done as follows:

$$\rho_{ki} = \sqrt{\rho_{ki-1}^2 + d^2 - 2\rho_{ki-1} d \cos\left(\frac{\pi}{2} + \theta_{ki-1}\right)} \quad (14)$$

$$\theta_{ki} = -\frac{\pi}{2} + \cos^{-1}\left(\frac{d^2 + \rho_{ki}^2 - \rho_{ki-1}^2}{2\rho_{ki-1} d}\right), i = 2, \ldots, N \quad (15)$$

These equations are employed in Eq. (5) to estimate simultaneously range and bearing of the objects.

In the following section, we summarize the proposed algorithm for fast localization of underwater acoustic sources using a wide-band transmitter to receive the signals at different frequencies, then the coherent signal subspace can be applied to decorrelate the source signals.

4. Proposed algorithm for fast localization of underwater acoustic sources

We use spatial smoothing to deal with narrow band correlated signal, we divide the array into L_s overlapping subarrays. The spatially smoothed covariance matrix is the average of the subarray covariances [33]. The step-by-step proposed algorithm for fast localization of underwater acoustic sources is given as following:

Algorithm 1 Proposed Algorithm for Bearing and Range Estimation of Buried Objects

1. use the beamformer method to find an initial estimate of $\widehat{\theta_{0k}}$, where $k=1,\dots,P_0$, with $P_0 \leq P$.
2. compute the initial values of $\rho_k = \frac{X}{\cos(\theta_k)}$ for $k=1,\dots,P_0$, where $X=h+y_c$ represents the distance between the receiver and the bottom of the tank (seabed),
3. fill the transfer matrix, $\widehat{\mathbf{A}}(f) = [\mathbf{a}(f,\theta_1,\rho_1),\mathbf{a}(f,\theta_2,\rho_2),\dots,\mathbf{a}(f,\theta_K,\rho_K)]$, where each source steering vector is filled using Eq. (11),
4. estimate the spectral matrix $\mathbf{\Gamma}(f) = E[\mathbf{r}(f)\mathbf{r}^+(f)] = \frac{1}{L_r}\sum_{l=1}^{L_r}\mathbf{r}_l(f)\mathbf{r}_l^+(f)$, where L_l is the realization number,
5. estimate noise covariance matrix $\Gamma_b(f) = E[\mathbf{b}(f)\mathbf{b}^+(f)]$. When the noise is white noise, that is, estimate noise variance $\sigma^2(f) = \frac{1}{N-P_0}\sum_{i=P_0+1}^{N}\lambda_i$, where λ_i is the ith eigenvalue of $\mathbf{\Gamma}(f)$. Then we calculate $\mathbf{\Gamma}_b(f)=\sigma^2\mathbf{I}$, where $\mathbf{I}$ is the identify matrix,
6. calculate the spectral matrix of the signals reflected on the objects by $\mathbf{\Gamma}_s(f) = \left(\widehat{\mathbf{A}}^+(f)\widehat{\mathbf{A}}(f)\right)^{-1}$ $\widehat{\mathbf{A}}^+(f)\times(\mathbf{\Gamma}(f)-\mathbf{\Gamma}_n(f))\widehat{\mathbf{A}}(f)\left(\widehat{\mathbf{A}}^+(f)\widehat{\mathbf{A}}(f)\right)^{-1}$,
7. compute the average of the spectral matrices $\overline{\mathbf{\Gamma}}_s(f) = \frac{1}{L_s}\sum_{S=1}^{L_S}\mathbf{\Gamma}_s(f)$, where L_s represents the number of subarrays, then calculate $\overline{\mathbf{V}}_s(f)$ by SVD,
8. calculate the spatial spectrum of the *MUSIC* method for bearing and range object estimation:

$$MUSIC = \frac{1}{||\mathbf{a}(f_0,\theta_k,\rho_k)^+\overline{\mathbf{V}}_b(f_0)\overline{\mathbf{V}}_b^+(f_0)\mathbf{a}(f_0,\theta_k,\rho_k)||}, \tag{16}$$

where $\overline{\mathbf{V}}_b$ is the eigenvector matrix of noise subspace associated with the $(N-P)$ smallest eigenvalues.

Inverse Power method can be used to estimate the noise power $\lambda_N(f) = \sigma^2(f)$. This variance can be used at step 5 in Alg. 1 for estimating the noise variance in case of white noise. The principle of this method is recalled in the following, using the maximum norm.

1. Let $\mathbf{q}_o$ a complex vector of N elements, $\|\mathbf{q}_o\|_\infty = 1$;
2. For $l = 1, 2, 3, \ldots$;
3. Calculate $\mathbf{x}_l : \boldsymbol{\Gamma}\mathbf{x}_l = \mathbf{q}_{l-1}$

   $\mu^l = \| \mathbf{x}_l\|_\infty$

   $\mathbf{q}_l = \frac{\mathbf{x}_l}{\mu^l}$;

It is shown that: $\lim_{l\to\infty}\mu^l = \frac{1}{\lambda_N}$, where λ_N is the smallest eigenvalue of $\boldsymbol{\Gamma}$.

In the high resolution noise subspace based methods, the DOA's are given by the local maximum points of a cost function, for example Eq. (16) of MUSIC. $\mathbf{V}_b(f)$ is the orthogonal projector onto the noise subspace given by the eigenvectors associated with the smallest eigenvalues of the covariance matrix of the received data. This requires an enormous computational load which limits its use for tracking by SVD [34]. In many applications, only a few eigenvectors are required. Since the number of sensor N is often larger than the number of sources P. It means that the vector dimension of noise subspace is larger than signal subspace. It is more efficient to work with the lower dimensional signal subspace than with the noise subspace. That is to say, it is not necessary to obtain $\mathbf{V}_b(f)$ exactly. We can calculate signal subspace $\mathbf{V}_s(f) = [\mathbf{v}_1(f), \mathbf{v}_2(f), \ldots, \mathbf{v}_P(f)]$ whose columns are the P orthonormal basis vectors. The projector onto the noise subspace spanned by the $(N-P)$ eigenvectors associated with the $(N-P)$ smallest eigenvalues is $\mathbf{V}_b(f)\mathbf{V}_b^+(f)$, given by:

$$\mathbf{V}_b(f)\mathbf{V}_b^+(f) = \mathbf{I} - \mathbf{V}_s(f)\mathbf{V}_s^+(f) \tag{17}$$

On the other hand, the additive noise is assumed to be white. But in practice, the noise is not always spatially white noise. In generally, the noise is correlated or unknown.

So in the next two sections, we will introduce the algorithms to replace SVD in MUSIC for reduce computation times and propose a new algorithm for estimating the spectral matrix of an unknown limited length spatially correlated noise.

5. MUSIC without eigendecomposition

In this section, we propose the noneigenvector versions of "MUSIC" to replace SVD to accelerate computation times.

5.1. Fixed point algorithm

One way to compute the P orthonormal basis vectors is to use Gram-Schmidt method. The eigenvector with dominant eigenvalue will be measured first. Similarly, all the remaining $P-1$ basis vectors will be measured one by one in a reducing order of dominance. The previously measured $(p-1)^{th}$ basis vectors will be utilized to find the pth basis vector. The algorithm for p^{th} basis vector will converge when the new value $\mathbf{v}_p^+$ and old value $\mathbf{v}_p$ are such that $\mathbf{v}_p^+\mathbf{v}_p = 1$. It is usually economical to use a finite tolerance error to satisfy the convergence criterion $\left|\left|\mathbf{v}_p^+\mathbf{v}_p - 1\right|\right| < \eta$ where η is a prior fixed threshold. The proposed algorithm is given as follows:

Algorithm 2 Fixed Point Algorithm

1. Choose P, the number of principal axes or eigenvectors required to estimate. Consider covariance matrix $\mathbf{\Gamma}$ and set $p \leftarrow 1$.
2. Initialize eigenvector $\mathbf{v}_p$ of size $d \times 1$, e.g. randomly;
3. while $\left|\left|\mathbf{v}_p^H\mathbf{v}_p - 1\right|\right| < \eta$
   a. Update $\mathbf{v}_p$ as $\mathbf{v}_p \leftarrow \mathbf{\Gamma}\mathbf{v}_p$;
   b. Do the Gram-Schmidt orthogonalization process $\mathbf{v}_p \leftarrow \mathbf{v}_p - \sum_{j=1}^{j=p-1}\left(\mathbf{v}_p^T\mathbf{v}_j\right)\mathbf{v}_j$;
   c. Normalize v_p by dividing it by its norm: $\mathbf{v}_p \leftarrow \frac{\mathbf{v}_p}{||\mathbf{v}_p||}$.
4. Increment counter $p \leftarrow p+1$ and go to step 2 until p equals K.

5.2. Projection approximation subspace tracking (PAST) algorithm

Suppose that we have an estimation of the signal subspace $\mathbf{W}(t)$ where each column is an eigenvector. The Linear Principal Analysis Criterion gives the definition of the scalar cost function $J(\mathbf{W}(t))$.

$$J(\mathbf{W}(t)) = E\left\{||\mathbf{r}(t) - \mathbf{W}(t)\mathbf{W}^+(t)\mathbf{r}(t)||^2\right\} \tag{18}$$

where $\mathbf{W}(t)\mathbf{W}^+(t)\mathbf{r}(t)$ is the projection of $\mathbf{r}(t)$ into the subspace $\mathbf{W}(t)$. The error surface of the function has several local minimal and one global minimum. When $\mathbf{W}(t)$ is equal to a basis for the signal subspace, $J(\mathbf{W}(t))$ has a global minimum which can estimate the signal subspace by Eq. (18). Note that $\mathbf{W}(t)$ is not equal to the signal subspace itself, but merely provides a possible basis. If $\mathbf{W}(t)$ is a signal column vector, it does indeed become equal to the Principal Component (dominant eigenvector) under minimization.

The cost function $J(\mathbf{W}(t))$ can be minimized by the application of a gradient-descent technique or recursive least squares variant. We can replace the expectation operator in Eq. (18) by an exponentially weighted sum over n samples. The estimation is given as follows:

$$\widehat{J}(J(\mathbf{W}(t))) = \sum_{t=1}^{n}\beta^{n-t}||\mathbf{r}(t) - \mathbf{W}(t)\mathbf{W}(t)^+\mathbf{r}(t)||^2 \tag{19}$$

where β is the forgetting factor ($0<\beta<1$). The forgetting factor allows the subspace estimation to track geostationary signal over time.

We get another cost function by approximating $\mathbf{W}(t)^{+}\mathbf{r}(t)$ with $\mathbf{W}(t-1)^{+}\mathbf{r}(t)$, using the previous value of $\mathbf{W}(t)^{+}$ in the iteration, giving:

$$\hat{J}'(\mathbf{W}(t)) = \sum_{t=1}^{n} \beta^{n-t} \left\| \mathbf{r}(t) - \mathbf{W}(t)\mathbf{W}(t-1)^{+}\mathbf{r}(t) \right\|^{2} \tag{20}$$

This function resembles the cost function used to define a recursive least squares (RLS) filter:

$$C(\mathbf{W}(t)) = \sum_{t=1}^{n} \beta^{n-t} |\mathbf{e}(t)|^{2} \tag{21}$$

where $\mathbf{e}(t)$ is the error signal. The error signal is the difference between the "desired" signal $\mathbf{r}(t)$ and its projection into the subspace $\mathbf{W}(t)\mathbf{W}(t-1)^{+}\mathbf{r}(t)$. Consequently, the PAST algorithm may be summarized by the following equations:

Algorithm 3 PAST Algorithm

1. Initialization:

   $\mathbf{W}(0)$ and $\mathbf{P}(0)$

2. for t = 1,2…

   $\mathbf{y}(t) = \mathbf{W}^{+}(t-1)\mathbf{r}(t)$

   $\mathbf{h}(t) = \mathbf{P}(t-1)^{*}\mathbf{y}(t)$

   $\mathbf{g}(t) = \frac{\mathbf{h}(t)}{(\beta + \mathbf{y}(t))\mathbf{h}(t)}$

   $\mathbf{P}(t) = \frac{1}{\beta}\mathrm{Tri}\{\mathbf{P}(t-1) - \mathbf{g}(t)\mathbf{r}^{+}(t)\}$

   $\mathbf{e}(t) = \mathbf{r}(t) - \mathbf{W}(t-1)\mathbf{y}(t)$

   $\mathbf{W}(t) = \mathbf{W}(t-1) + \mathbf{e}(t)\mathbf{g}^{+}(t)$

   end

The operator *Tri* indicates that only the upper (or lower) triangular part of the matrix is calculated and its Hermitian transposed version is copied to the another lower (or upper) triangular part.

5.3. Projection approximation subspace tracking with deflation (PASTD) algorithm

The PAST algorithm provides a method to estimate only a basis for the dominant subspace. The exact eigenvectors (singular vector) are not calculated unless $\mathbf{W}(t)$ is a column vector in which case only the dominant eigenvector (principal component) is estimated. We present a second subspace tracking algorithm - PAST with deflation (PASTD) which is derived from the PAST approach. The PASTD algorithm is based on the deflation technique which is the sequential estimation of the principal components. According to the Karhunen-Loève expansion

$$\mathbf{r}(t) = \sum_{i=1}^{N} \mathbf{W}_i^+(t)\mathbf{r}(t)\mathbf{W}_i(t) \tag{22}$$

tells that any $\mathbf{r}(t)$ may be expressed as a linear combination of the eigenvector of the correlation matrix.

The first step of PASTD is to update the most dominant eigenvector by applying PAST with $i = 1$, then the contribution of the dominant eigenvector in Eq. (22) is removed by subtraction. So the second dominant eigenvector becomes the most dominant and can be extracted in the same way. Then we repeat the procedure until all desired eigencomponents are estimated. This iterative process is called **deflation**. So the algorithm may be summarized as follows:

Algorithm 4 PASTD Algorithm

1. Initialization:

$\mathbf{W}_i(0)$ and $d_i(0)$

2. for n = 1,2…

$\mathbf{r}_1(t) = \mathbf{r}(t)$

for i = 1,2,…,P $\mathbf{y}_i(t) = \mathbf{W}_i^+(t-1)\mathbf{r}_i(t)$

$d_i(t) = \beta d_i(t-1) + |\mathbf{y}_i(t)|^2$

$\mathbf{e}_i(t) = \mathbf{r}_i(t) - \mathbf{W}_i(t-1)\mathbf{y}_i(t)$

$\mathbf{W}_i(t) = \mathbf{W}_i(t-1) + \mathbf{e}_i(t)\mathbf{y}_i^*(t)/d_i(t)$

$\mathbf{r}_{i+1}(t) = \mathbf{r}_i(t) - \mathbf{W}_i(t)\mathbf{y}_i(t)$

end

end

where estimates are made of P eigenvectors with the largest eigenvalues. In practice since $P << N$, this indicates an important optimization compared to the eigendecomposition or SVD. The eigenvector projection estimates $\mathbf{W}_i$ are initialized to the columns of some nonzero orthogonal matrix. $d_i(t)$ is initialized to arbitrary nonzero constants. When PASTD has converged, the $\mathbf{W}_i(t)$ will contain estimates of the eigenvector of the correlation matrix of the data in $\mathbf{r}(t)$. The corresponding eigenvalues may be calculated by multiplying the $d_i(t)$ by $\frac{1-\beta}{\beta}$.

5.4. Orthogonal projection approximation subspace tracking (OPAST) algorithm

The OPAST algorithm is the modification of PAST. The weight matrix $\mathbf{W}(t)$ is forced to be orthonormal to each iteration. So we can get:

$$\mathbf{W}(t) = \mathbf{W}(t)(\mathbf{W}^+(t)\mathbf{W}(t))^{-1/2} \tag{23}$$

where $(\mathbf{W}^{+}(t)\mathbf{W}(t))^{-1/2}$ denotes an inverse square root of $(\mathbf{W}^{+}(t)\mathbf{W}(t))$. $(\mathbf{W}^{+}(t)\mathbf{W}(t))^{-1/2}$ can be calculated by using the updating equation of $\mathbf{W}(t)$. Note that $\mathbf{W}(t-1)$ is now an orthonormal matrix, we have

$$\mathbf{W}^{+}(t)\mathbf{W}(t) = \mathbf{I} + \|\mathbf{p}(t)\|^{2}\mathbf{q}(t)\mathbf{q}^{+}(t) \tag{24}$$

where $\mathbf{I}$ is the identity matrix, $\mathbf{W}^{+}(t-1)\mathbf{p}(t)=0$ and $\mathbf{r} =^{def} \|\mathbf{p}(t)\|\mathbf{q}(t)$. Thus

$$\begin{aligned}(\mathbf{W}^{+}(t)\mathbf{W}(t))^{-1/2} &= \mathbf{I} + \frac{1}{\|\mathbf{r}\|^{2}}\left(\frac{1}{\sqrt{1+\|\mathbf{r}\|^{2}}} - 1\right)\mathbf{x}\mathbf{x}^{H} \\ &= \mathbf{I} + \tau(t)\mathbf{q}(t)\mathbf{q}^{+}(t)\end{aligned} \tag{25}$$

where

$$\tau(t) \overset{def}{=} \frac{1}{\|\mathbf{q}(t)\|^{2}}\left(\frac{1}{\sqrt{1+\|\mathbf{p}(t)\|^{2}\|\mathbf{q}(t)\|^{2}}} - 1\right) \tag{26}$$

Using Eqs. (23) and (26), and the updating equation of $\mathbf{W}(t)$, we obtain

$$\begin{aligned}\mathbf{W}(t) &= (\mathbf{W}(t-1) + \mathbf{p}(t)\mathbf{q}^{+}(t))(\mathbf{I} + \tau(t)\mathbf{q}(t)\mathbf{q}^{+}(t)) \\ &= \mathbf{W}(t-1) + \mathbf{p}'(t)\mathbf{q}^{+}(t)\end{aligned} \tag{27}$$

where $\mathbf{p}'(t) = \tau(t)\mathbf{W}(t-1)\mathbf{q}(t) + (1+\tau(t)\|\mathbf{q}(t)\|^{2})\mathbf{p}(t)$. Thus, the OPAST algorithm can be written as the PAST (see Algorithm 3):

Algorithm 5 OPAST Algorithm

$\mathbf{W}(t) = \mathbf{W}(t-1) + \mathbf{p}'(t)\mathbf{q}^{+}(t)$

$\tau(t) = \frac{1}{\|\mathbf{q}(t)\|^{2}}\left(\frac{1}{\sqrt{1+\|\mathbf{p}(t)\|^{2}\|\mathbf{q}(t)\|^{2}}} - 1\right)$

$\mathbf{W}(t) = \tau(t)\mathbf{W}(t-1)\mathbf{q}(t) + (1+\tau(t)\|\mathbf{q}(t)\|^{2})\mathbf{p}(t)$

6. Simulations results

6.1. Complexity

The traditional MUSIC method estimate the noise subspace eigenvectors by SVD. From the computational point of view, the well-known SVD method is the cyclic Jacobi's method which requires around N^3 computations. The computational complexity of fixed-point algorithm, PAST, PASTD and OPAST is (NP^2+N^2P), $3NP+O(P^2)$, $4NP+O(P)$ and $4NP+O(P^2)$ respectively. If the number of sensors N is larger compared to the number of objects P, the computational complexity can be estimated to be around N^2P for fixed-point algorithm, $3NP$ for PAST, $3NP$

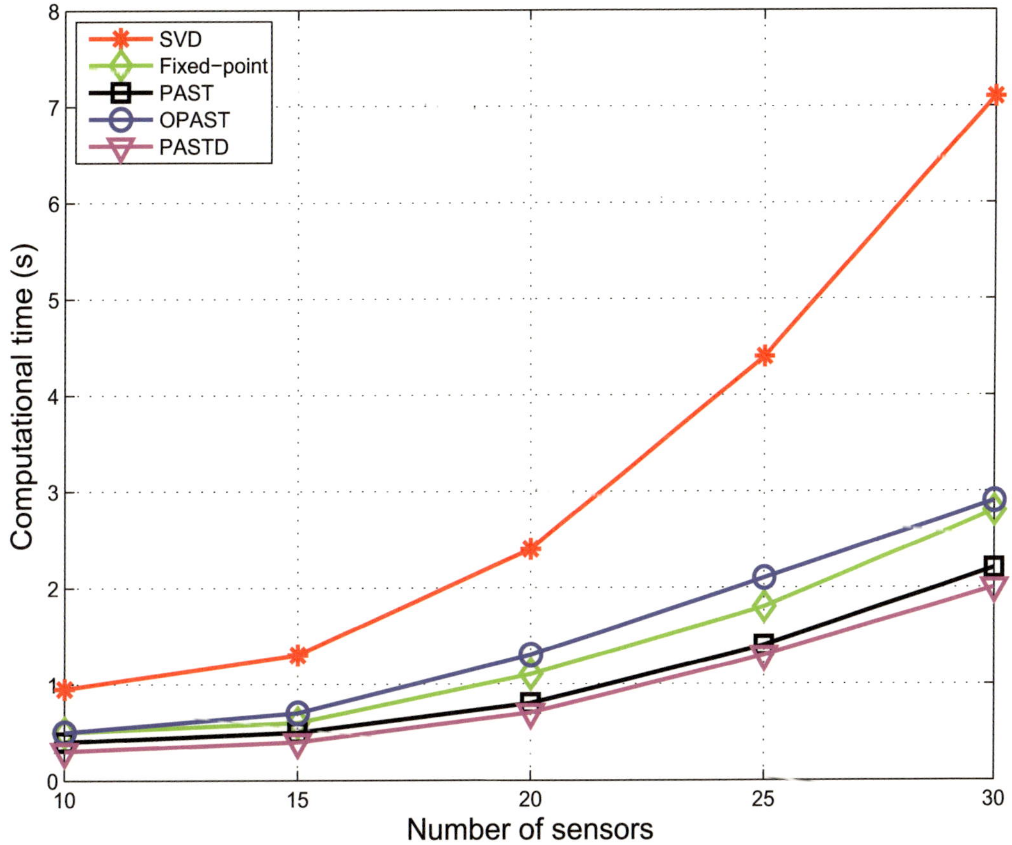

Figure 3. Computational times, SVD (red), Fixed–point (green), PAST (Black), OPAST(blue) and PASTD (pink).

	10	15	20	25	30
Time SVD (second)	0.95	1.3	2.4	4.4	7.1
Time fixed point (second)	0.5	0.6	1.1	1.8	2.8
Time PAST (second)	0.4	0.5	0.8	1.4	2.2
Time PASTD (second)	0.3	0.4	0.7	1.3	2.0
Time OPAST (second)	0.5	0.7	1.3	2.1	2.9
Ratio SVD/fixed point	1.9	2.2	2.2	2.4	2.5
Ratio SVD/PAST	2.4	2.6	3.0	3.1	3.2
Ratio SVD/PASTD	3.2	3.3	3.4	3.4	3.6
Ratio SVD/OPAST	1.9	1.9	1.8	2.1	2.4

Table 1. Computational time needed to run MUSIC for various numbers of sensors.

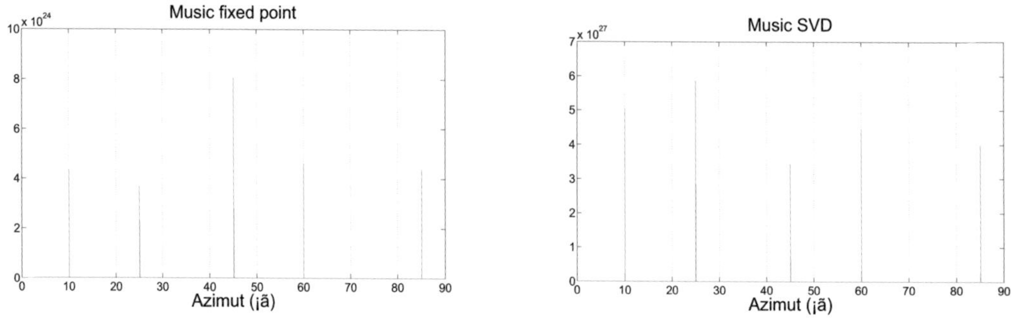

(a) (b)

Figure 4. (a) Pseudospectrum of MUSIC obtained using fixed point, (b) pseudospectrum of MUSIC obtained using SVD.

for PASTD and 4*NP* for OPAST. Several experiments are carried out with various numbers of sensors, to study the computational load of the proposed algorithm with $SNR = 0$ dB and $P = 8$ sources. DOA values are: 5°, 10°, 20°, 25°, 35°, 40°, 50° and 55°.

The number of realizations is 1000, and the number of observations is 1000. Choosing a number of snapshots equal to 100, such as in [14, 21, 22, 29, 30], does not change the results. The mean computational load is then up to 2.5 times less with fixed point algorithm than with SVD (see **Figure 3** and **Table 1**, N = 10 up to 30). Both versions of MUSIC provide the same results (see **Figure 4**, take fixed-point algorithm for example).

7. Experimental setup

The studied signals are recorded during an underwater acoustic experiment in order to estimate the developed method performance. The experiment is carried out in an acoustic tank under the conditions similar to those in a marine environment. The bottom of the tank is filled with sand. The experimental device is presented in **Figure 5**. The tank is topped by two mobile carriages. The first carriage supports a transducer issuer and the second supports a transducer receiver pilot by the computer.

Four couples of spherical and cylindrical shells (see **Figure 6**) are buried between 0 and 0.05 m under the sand. The considered objects have the following characteristics, where δ represents the distance between the two objects of the same couple and $\varnothing_a$ the outer radius (the inner radius $\varnothing_b = \varnothing_a - 0.001$ m):

1. the 1st couple (O_1, O_2): spherical shells, $\varnothing_a = 0.3$ m, $\delta = 0.33$ m, full of air,
2. the 2nd couple (O_3, O_4): cylindrical shells, $\varnothing_a = 0.01$ m, $\delta = 0.13$ m, full of air,
3. the 3rd couple (O_5, O_6): cylindrical shells, $\varnothing_a = 0.018$ m, $\delta = 0.16$ m, full of water,
4. the 4th couple (O_7, O_8): cylindrical shells, $\varnothing_a = 0.02$ m, $\delta = 0.06$ m, full of air,

The considered objects are made of dural aluminum with density $D_2 = 1800$ kg/m^3, the speed of the wave in the water c_1 is 1500 m/s and in the sediment c_2 is 1700 m/s, the longitudinal and

http://dx.doi.org/10.5772/intechopen.71272

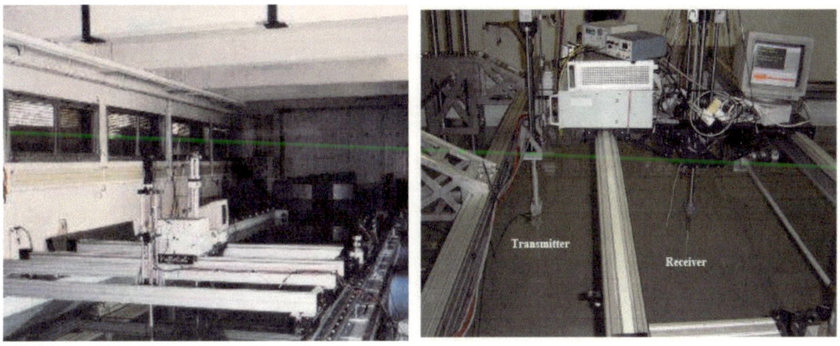

(a) (b)

Figure 5. Experimental setup: (a) Data acquisition system, (b) Experimental tank.

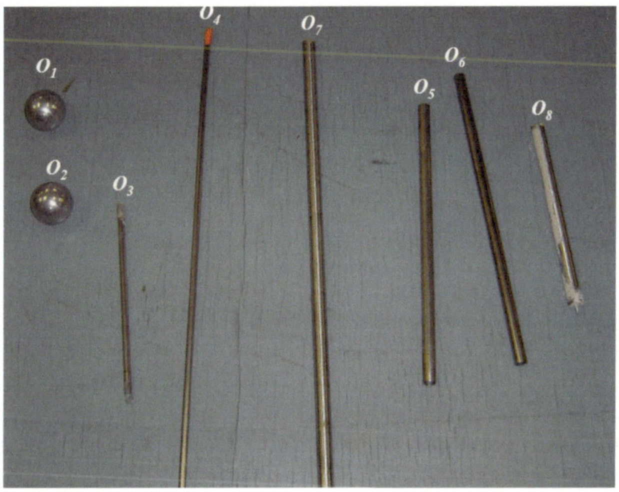

Figure 6. Experimental objects.

transverse-elastic wave velocities inside the shell medium are $c_l = 6300$ m/s and $c_t = 3200$ m/s, respectively. The speed of the wave in the water c_1 is 1500 m/s and in the sediment c_2 is 1700 m/s, The external fluid is water with density $D_1 = 1000$ kg/m^3 and the internal fluid is water or air with density $D_{3air} = 1.2 \cdot 10^{-6}$ kg/m^3 or $D_{3water} = 1000$ kg/m^3.

In addition to estimate the performance of the propose method, the signal source a spatially correlated noise is emitted with $K = 10$. The objective is to estimate the directions of arrival of the signals during the experiment. The signals are received on one uniform linear array. The observed signals come from various reflections on the objects being in the tank. Generally the aims of acousticians is the detection, localization and identification of these objects. In this experiment we have recorded the reflected signals by a single receiver. This receiver is moved along a straight line between position $X_{min} = 50$ mm and position $X_{max} = 150$ mm with a step of $\alpha = 1$ mm in order to create a uniform linear array. The experimental setup is shown in **Figure 7**. We have measured eight times $E_i(O_{ii}, O_{ii+1})$ with $i = 1, \ldots, 8$ and $ii = 1, 3, 5, 7$. At first, the receiver horizontal axis XX' is fixed at 0.2 m, we performed the experiments $E_1(O_1, O_2), \ldots, E_4(O_7, O_8)$

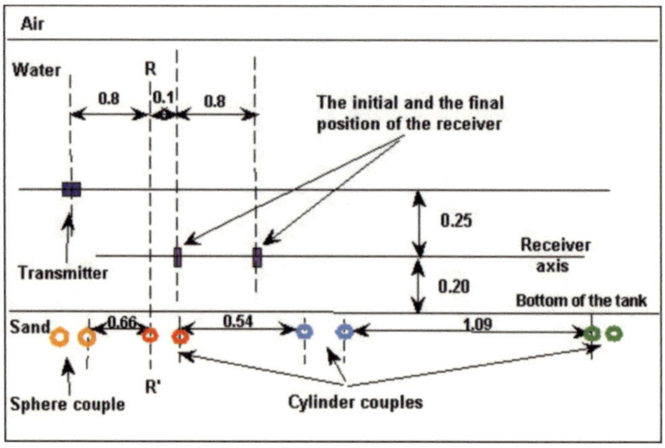

Figure 7. Experimental setup.

associated to the 1st, 2nd, 3rd and the 4th couple, respectively. Then we performed the other four experiments $E_5(O_1, O_2), \ldots, E_8(O_7, O_8)$ with XX' fixed at 0.4 m. RR' is the vertical axis which goes through the center of the first object of each couple. For each experiment, the transmitted signal had the following properties: pulse duration is 15 μs, the frequency band is, the frequency of the band is $[f_{min}=150, f_{max}=250]$ kHz and the center frequency f_0 is $f_0=200$ kHz. The sampling rate is 2 MHz. The duration of the received signal was 700 μs. The variance of Gaussian white noise σ^2 is 100 and the angle of incidence θ_{inc} is 60°.

At each sensor, time-domain data corresponding only to target echoes are collected with signal to noise ratio equal to 20 dB. The typical sensor output signals recorded during one experiment is shown in **Figure 8**.

The proposed algorithms were applied on each experimental data set. Forming the directional vector by the model of acoustic diffusion appropriate to locate the objects and the spectral matrix of the simulated data. We use the focusing operator on the signals by dividing the

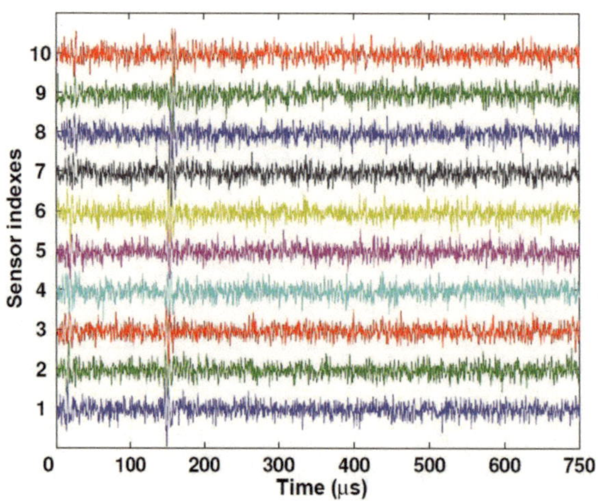

Figure 8. Example of observed signals during experiment Exp. 1.

http://dx.doi.org/10.5772/intechopen.71272

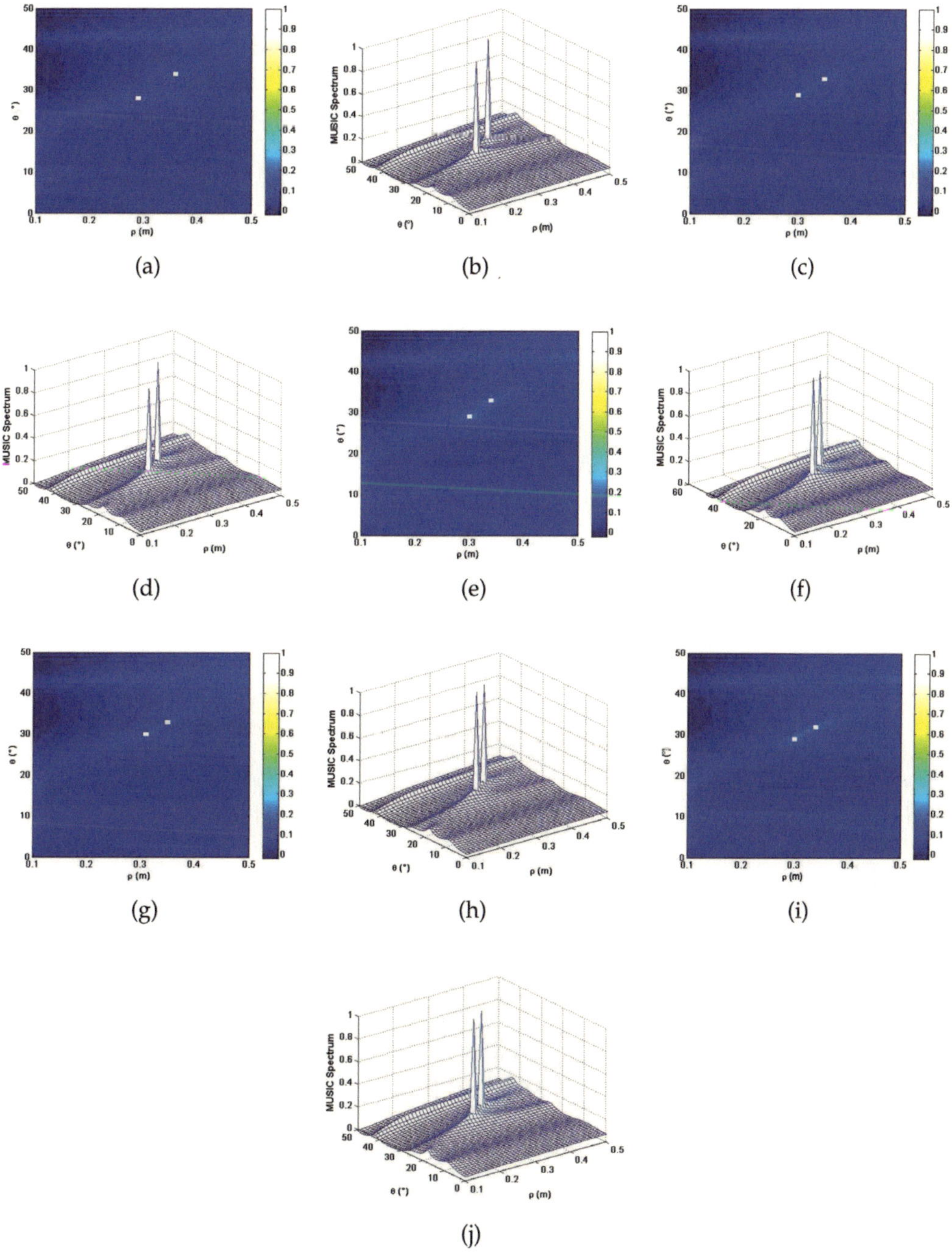

Figure 9. Example of object localization with different methods: (a)-(b) SVD, (c)-(d) Fixed-point, (e)-(f) PAST, (g)-(h) PASTD and (i)-(j) OPAST.

frequency band [150, 250] kHz in 11 frequencies and Alg. 2, Alg. 3, Alg. 4 and 5 to calculate the noise subspace, respectively. Finally we apply Eq. (16) to estimate DOA of objects and object-1st sensor distance.

As is shown in **Figure 9**, X axis is the object-1st sensor distance ρ, Y axis is the DOA of object-1st sensor θ. The white points and the peak positions, which present the maximum values, correspond to the coordinate of 2 objects (29°, 0.31 m) and (33°, 0.34 m). The bearings and the ranges of buried objects are (28.1°, 0.298 m) and (33.9°, 0.361 m) if we use conventional SVD (see **Figure 9(a)** and **(b)**). The results of the proposed algorithms are (29.5°, 0.301 m) and (33.3°, 0.351 m) for fixed-point algorithm (see **Figure 9(c)** and **(d)**), (29.2°, 0.312 m) and (32.8°, 0.343 m) for PAST algorithm (see **Figure 9(e)** and **(f)**), (29.8°, 0.313 m) and (33.3°, 0.355 m) for PASTD algorithm (see **Figure 9(g)** and **(h)**) and (29.3°, 0.306 m) and (32.5°, 0.334 m) for OPAST algorithm (see **Figure 9(i)** and **(j)**).

We have done statistical study in order to a posteriori verify the quality of estimation of the proposed method. Standard Deviation (Std) is defined as follows:

$$std = \sqrt{\frac{1}{N}\sum_{i=1}^{N}\left(X_{iexp} - X_{iest}\right)^2} \tag{28}$$

where X_{iexp} (respectively X_{iest}) represents the ith expected (respectively the ith estimated) value of θ or ρ. Standard deviation of the bearing and the range estimation at different signal-to-noise ratio (SNRs) (from −10 to 20 dB) are given in **Figure 10**.

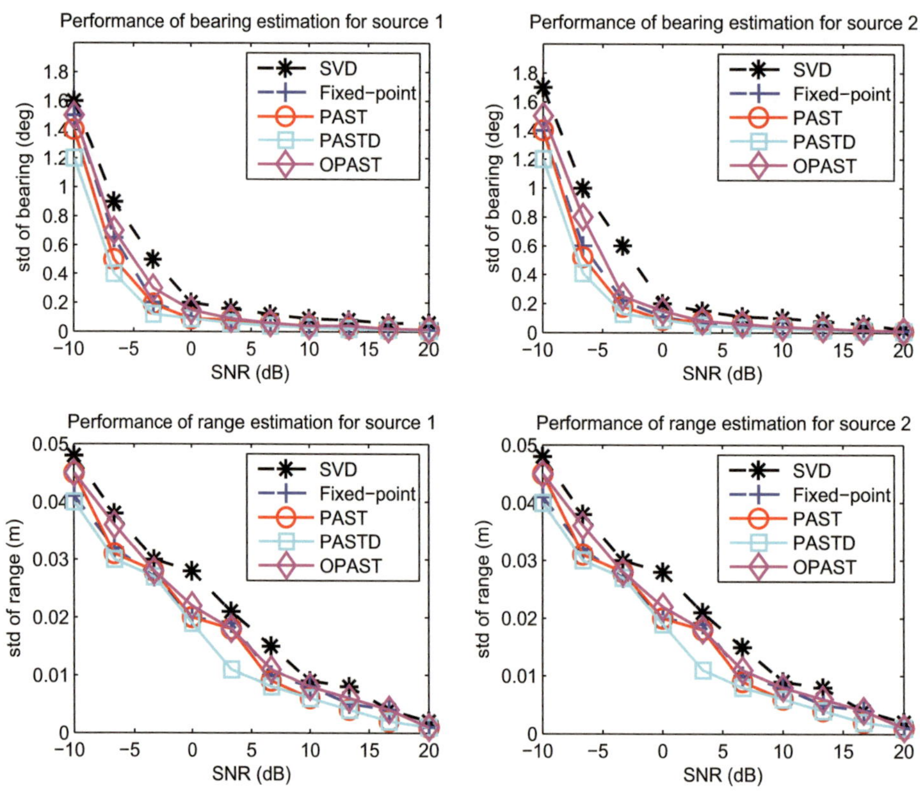

Figure 10. Standard deviation versus SNR of the bearing and the range estimation.

8. Conclusion

The main target of array processing is the estimation of the parameters: DOA of objects and the objects-sensors distance. In this chapter, we have proposed a new fast localization algorithm to estimate both the ranges and the bearings of buried sources underwater acoustic in presence of correlated noise. This algorithm takes into account both the reflection and refraction of water-sediment interface. We develop fixed point algorithm in MUSIC instead of SVD to keep the small computational time load. A new focusing operator is proposed to estimate the coherent signal subspace. Some simulations have been done to test our method. We compare the computation time of MUSIC with SVD and fixed point, it shows that fixed point is faster than SVD. The proposed method performance was investigated through scaled tank tests associated with some cylindrical and spherical shells buried in an homogenous fine sand. The obtained results are promising and the *RE* calculated between the expected and the estimated bearings and ranges errors is weep.

Author details

Salah Bourennane* and Caroline Fossati

*Address all correspondence to: salah.bourennane@fresnel.fr

Aix Marseille Univ, CNRS, Centrale Marseille, Institut Fresnel, France

References

[1] Bourennane S, Frikel M. Localization of wideband sources with estimation of an antenna shape. In: Proc. IEEE Workshop on Statistical Sig. Array Proc.; 1996. pp. 97-100

[2] Bermani E, Boni A, Caorsi S, Massa A. An innovative real-time technique for buried object detection. IEEE Transactions on Geoscience and Remote Sensing. 2003;**41**:927-931

[3] Caorsi S, Anguita D, Bermani E, Boni A, Donelli M, Massa A. A comparative study of NN and SVM-based electromagnetic inverse scattering approaches to on-line detection of buried objects. Journal of the Applied Computational Electromagnetics Society. Jul. 2003;**18**:1-11

[4] Massa A, Boni A, Donelli M. A classification approach based on SVM for electromagnetic subsurface sensing. IEEE Transactions on Geoscience and Remote Sensing. Sept. 2005;**43**(9): 2084-2093

[5] Zhang B, O'Neill K, Kong JA, Grzegorczyk TM. Support vector machine and neural network classification of metallic objects using coefficients of the spheroidal MQS response modes. IEEE Transactions on Geoscience and Remote Sensing. Jan. 2008;**46**(1):159-171

[6] Ho KC, Carin L, Gader PD, Wilson JN. An investigation of using the spectral characteristics from ground penetrating radar for landmine/clutter discrimination. IEEE Transactions on Geoscience and Remote Sensing. Apr. 2008;**46**(4):1177-1191

[7] Guillermin R, Lasaygues P, Sessarego JP, Wirgin A. Characterization of buried objects by a discretized domain integral equation inversion method using born approximation. In: Proc. 5th European Conference of Underwater Acoustics. Vol. 2; Jul. 10-13, 2000. pp. 863-868

[8] Hetet A, Amate M, Zerr B. SAS processing results for the detection of buried objects with a ship-mounted sonar. In: the 7th European Conference on Underwater Acoustics (ECUA '04); Delft, The Netherlands; July. 2004. pp. 1127-1132

[9] Kumareasan R, Tufts DW. Estimating the angles of arrival of multiple source plane waves. IEEE Transactions on Aerospace and Electronic Systems. 1983;**19**:134-139

[10] Baggeroer AB, Kuperman WA, Mikhalevsky PN. An overview of matched field methods in ocean acoustics. IEEE Journal of Oceanic Engineering. 1993;**18**(4):401-424

[11] Sahin A, Miller EL. Object-based localization of buried objects using high resolution array processing techniques. Proc. SPIE Aerosense. May 1996;**2765**

[12] Fawcett JA, Lim R. Evaluation of the integrals of target/seabed scattering using the method of complex images. The Journal of the Acoustical Society of America. Sep. 2003;**114**(3):1406-1415

[13] Fawcett JA, Fox WLJ, Maguer A. Modeling of scattering by objects on the seabed. The Journal of the Acoustical Society of America. Dec. 1998;**104**(6):3296-3304

[14] Bourennane S, Bendjama A. Locating wide band acoustic sources using higher-order statistics. Applied Acoustics. 2002;**63**:235-251

[15] Saidi Z, Bourennane S. Cumulant-based coherent signal subspace method for bearing and range estimation. EURASIP Journal on Advances in Signal Processing. 2007. p. 9. Article ID 84576. Doi:10.1155/2007/84576

[16] Doolittle RD, Uberall H. Sound scattering by elastic cylindrical shells. The Journal of the Acoustical Society of America. 1966;**39**(2)

[17] Ye Z. Recent developments in underwater acoustics: Acoustic scattering from single and multiple bodies. Proceedings of the National Science Council, Republic of China, Part A: Physical Science and Engineering. 2001;**25**(3):137-150

[18] Lim R. Acoustic scattering by a partially buried three dimensional elastic obstacle. The Journal of the Acoustical Society of America. 1998;**104**(2):769-782

[19] Tesei A, Maguer A, Fox WLJ, Schmidt H. Measurements and modeling of acoustic scattering from partially and completely buried spherical shells. The Journal of the Acoustical Society of America. Nov. 2002;**112**(5):1817-1830

[20] Bourennane S, Han D, Fossati C. Fast coherent signal subspace-based method for bearing and range of buried objects estimation in the presence of colored noise. IEEE Journal of Selected Topics in Applied Earth Observations and Remote Sensing, 2009;**2**:329-338

[21] Tayem N, Kwon HM. Azimuth and elevation angle estimation with no failure and no eigen decomposition. EURASIP Signal Processing. Jan. 2006;**86**(1):8-16

[22] Bourennane S, Fossati C, Marot J. About noneigenvector source localization methods. EURASIP Journal on Advances in Signal Processing. 2008 Article ID 480835

[23] Hyvärinen A, Oja E. A fast fixed-point algorithm for independent component analysis. Neural Computation. Oct. 1997;**9**(7):1483-1492

[24] Yang B. Projection approximation subspace tracking. IEEE Transactions on Signal Processing. Jan. 1995;**44**(1):95-107

[25] Yang B. Asymptotic convergence analysis of the projection approximation subspace tracking algorithms. Signal Processing. Apr. 1996;**50**:123-136

[26] Yang B. An extension of the PASTd algorithm to both rank and subspace tracking. IEEE Signal Processing Letters. Mar. 2000;**7**(3):60-62

[27] Golub GH, Van Loan CF. Matrix Computations. third ed. Baltimore and London: The Johns Hopkins University Press; 1996

[28] Hua Y, Xiang Y, Chen T, Abed-Meraim K, Miao Y. A new look at the power method for fast subspace tracking. Digital Signal Processing. Oct. 1999:297-314

[29] Abed-Meraim K, Chkeif A, Hua Y. Fast orthogonal PAST algorithm. IEEE Signal Processing Letters. 1995;**2**:179-182

[30] Badeau R, Abed-Meraim K, Richard G, David B. Sliding window orthonormal PAST algorithm. In: IEEE Int. Conf. on Acoustic, Speech and Signal Processing; 2003

[31] Gonen E, Mendel JM. Applications of cumulants to array processing - Part III: Blind beamforming for coherent signals. IEEE Transactions on Signal Processing. Sept. 1997;**45**(9): 2252-2264

[32] Mendel JM. Tutorial on higher-order statistics (spectra) in signal processing and system theory: Theoretical results and some applications. IEEE-Proceedings. Mar. 1991;**79**(3):278-305

[33] Villemin G, Fossati C, Bourennane S. Efficient time of arrival estimation in the presence of multipath propagation. The Journal of the Acoustical Society of America (JASA). Oct. 2013; **134**(4):301-306

[34] Villemin G, Fossati C, Bourennane S. Spatio-temporal-based joint range and angle estimation for wideband signals. EURASIP Journal on Advances in Signal Processing. 2013. Article ID 131

Routing Protocols for Underwater Acoustic Sensor Networks: A Survey from an Application Perspective

Qian Lu, Feng Liu, Ying Zhang and Shengming Jiang

Additional information is available at the end of the chapter

http://dx.doi.org/10.5772/intechopen.68900

Abstract

Underwater acoustic communications are different from terrestrial radio communications; acoustic channel is asymmetric and has large and variable end-to-end propagation delays, distance-dependent limited bandwidth, high bit error rates, and multi-path fading. Besides, nodes' mobility and limited battery power also cause problems for networking protocol design. Among them, routing in underwater acoustic networks is a challenging task, and many protocols have been proposed. In this chapter, we first classify the routing protocols according to application scenarios, which are classified according to the number of sinks that an underwater acoustic sensor network (UASN) may use, namely single-sink, multi-sink, and no-sink. We review some typical routing strategies proposed for these application scenarios, such as cross-layer and reinforcement learning as well as opportunistic routing. Finally, some remaining key issues are highlighted.

Keywords: underwater acoustic sensor networks (UASNs), application scenarios, routing protocol strategies

1. Introduction

Almost 71% of our plant is covered by oceans, and underwater networks become more and more important since they can be used for underwater information exchange, surveillance, ocean explorations and disaster prevention, etc., [1].

There are several challenges for underwater acoustic communication. First, because electromagnetic wave does not perform well in underwater environments due to the serious attenuation, acoustic communication is used as a major communication technology in underwater networks [2, 3]. The acoustic channel is asymmetric with large end-to-end propagation delays and limited bandwidth as well as high bit error rates due to multi-path fading. Second, underwater sensor nodes' mobility caused by ocean currents brings intermittent connective links. Third, underwater sensor nodes are prone to failure because of corrosion, fouling, and limited battery power that is available. Furthermore, the speed of the sound can change with water temperature, which leads to changes in the transmission path and may be the cause of the data not being forwarded to the sea surface on time. Therefore, the routing protocols available for terrestrial wireless sensor networks are not suitable for underwater acoustic sensor networks (UASNs), and many new routing protocols have been studied.

There are several surveys on routing protocols for UASNs reported in the literature, which provide overviews about the basic ideas of the related protocols mainly following the taxonomy for terrestrial wireless sensor networks. In Ref. [4], the authors investigated the characteristics and algorithms of routing protocols and classified them into the non–cross-layer design, the traditional cross-layer design, and the intelligent algorithm design. The work in Ref. [5] discussed the current state of the art on the UASN protocol design and provided a detailed overview on the current solutions for medium access control, network, and transport layer protocols. A review and comparison of different algorithms was proposed in Ref. [6] to fulfill different application requirements with dynamic environmental conditions. All these investigations show that it is almost impossible to conclude that any particular routing strategy can cost-effectively support all underwater applications because each of them has certain strengths and weaknesses and is only applicable to specific situations.

In this chapter, we conduct a review mainly from an application perspective. An optimal protocol design must take into account different favorable features available in different scenarios to maximize routing performance, especially the characteristics of various network topologies. Therefore, it is interesting to investigate how each protocol exploits these features to maximize protocol performance and the feasibility of the proposed schemes.

The rest of the chapter is organized as follows. In Section 2, we briefly introduce the characteristics of application scenarios and routing strategies. In Sections 3–5, we discuss typical routing protocols of different strategies in single-sink–, multi-sink–, as well as zero-sink–based UASNs. In Sections 6–8, we discuss typical routing protocols based on cross-layer and reinforcement learning as well as opportunistic routing protocols. Finally, in Section 9, we draw the conclusions.

2. Major application scenarios and routing strategies

2.1. Application scenarios

Most applications of UASNs are related to underwater collection, where the surface units are used to collect data transmitted by underwater sensor nodes. Therefore, application scenarios can be classified into single-sink, multi-sink, and zero-sink according to the number of sinks that a UASN uses. As shown in **Figure 1**, in a single-sink–based UASN, there is only one sink node which can be static or mobile. A static sink node is fixed on the surface (in some cases on

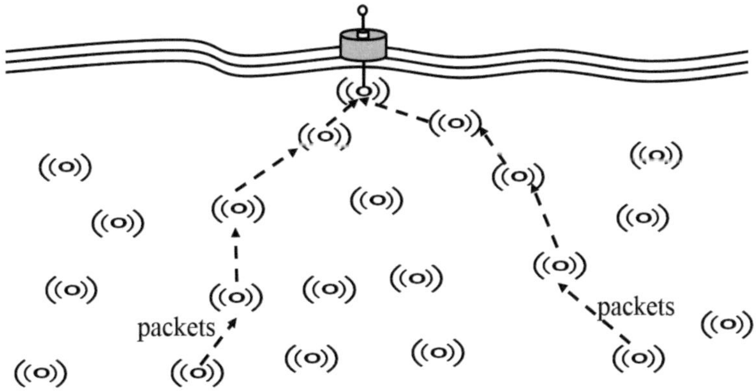

Figure 1. A static single-sink–based UASN.

the bottom) of the ocean as shown in **Figure 1** [1]. A mobile sink node moves to collect data from the sensor nodes deployed in the ocean as shown in **Figure 2**. The topology of a multi-sink–based UASN is shown in **Figure 3**, where two or more sink nodes are used to receive packets collected by sensor nodes [1]. A sensor node only needs to transmit the packets to one of the sink nodes close to the node. Sensor nodes in a multi-sink–based UASN can also be static or mobile. In a zero-sink–based UASN, several functionally identical autonomous underwater vehicles (AUVs) usually work as a team collaboratively, which requires communicating with each other, as illustrated in **Figure 4**.

Major factors that can affect the design of a routing protocol include the number of sink nodes and the topology of the corresponding UASN. Arranging multiple sink nodes in the network can improve the routing performance by shortening transmission path. A sensor node only needs to transmit the packets to the sink closer to it [1]. Deploying a UASN with static topology or a single-sink can simplify the design of a routing protocol. A zero-sink–based UASN may have several AUVs to work as a team, and it is more difficult to design routing protocols for this kind of UASN to achieve good performance.

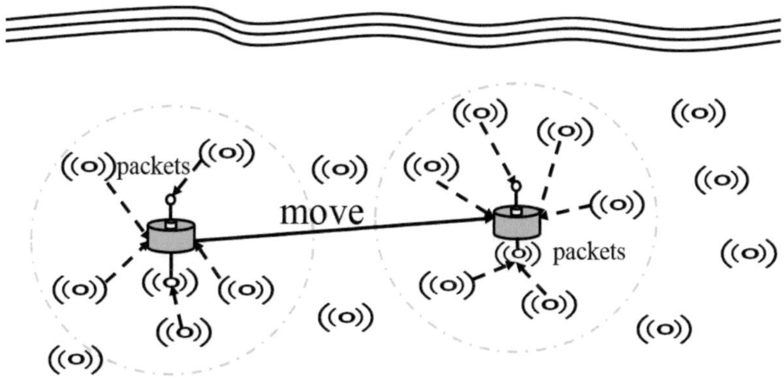

Figure 2. A mobile single-sink–based UASN.

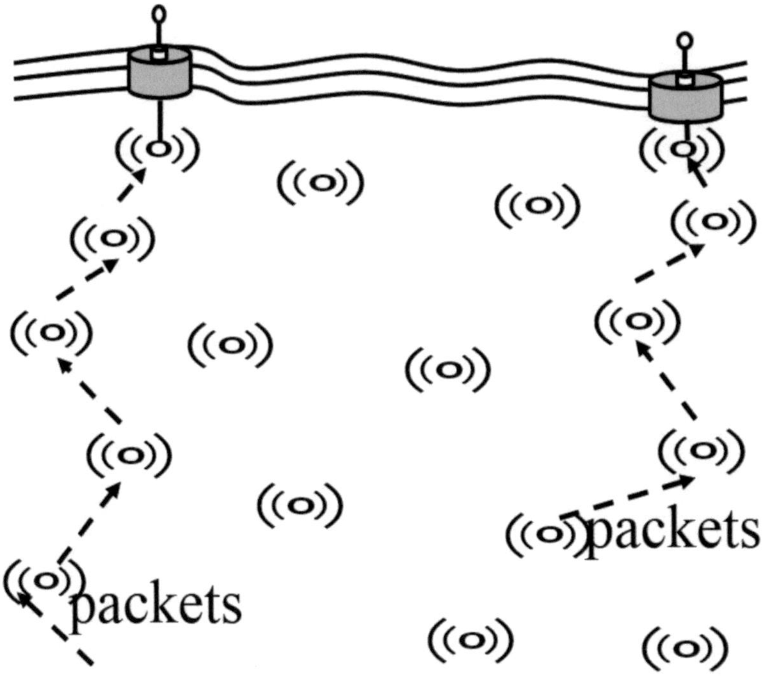

Figure 3. A multi-sink–based UASN.

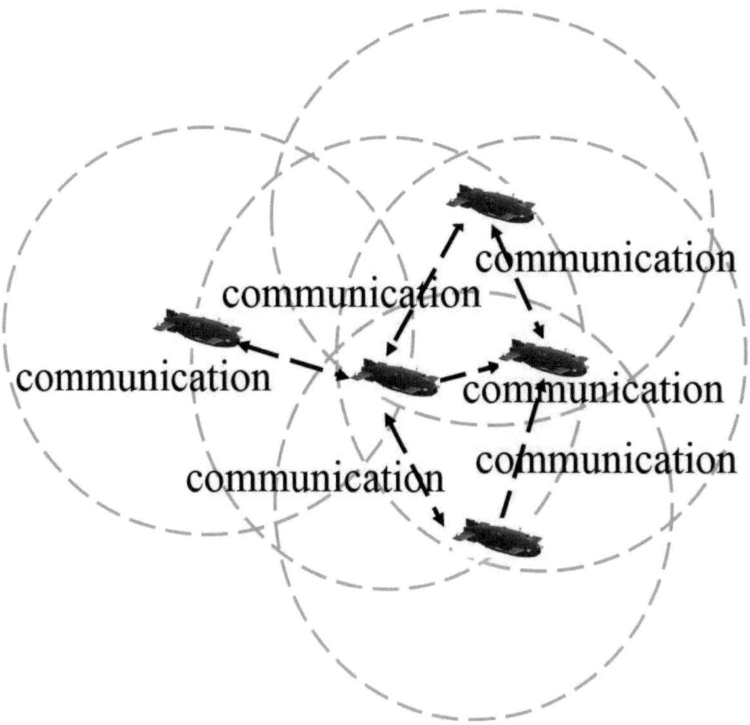

Figure 4. A zero-sink–based UASN.

2.2. Routing strategies

Table 1 lists routing strategies under reviewing.

Routing strategy	Characteristics	Application scenarios	Advantages	Disadvantages
Geo-based routing	It is based on GPS	All scenarios	Network architecture is simple	Positioning may be not accurate
Source routing	The transmission path is determined by the source node	Single-sink–based UASN with a static sink	It reduces the cost of route maintenance	It increases the packet overhead and routing cost
Hop-by-hop routing	The next-hop is determined by relay nodes	Except zero-sink–based UASN	Network is more scalable and flexible	The final path may be not optimal
Clustered routing	The nodes are usually divided into groups	Single-sink–based UASN	It has a good adaptability and can reduce data redundancy	It needs to select the cluster head
Cross-layer routing	It considers the characteristics of other layers	Single-sink–based UASN with a static sink and zero-sink–based UASN	It can minimize energy costs while maximizing the performance of the entire network	Routing protocol is more complex
Reinforcement learning scheme	Each node selects next hop via an iterative calculation of the reinforcement function	Single-sink–based UASN	It extends the lifetime of the network	More powerful nodes are required
Opportunistic routing	Each node selects several suitable neighbor nodes as a set of candidate-forwarding nodes	Multi-sink–based UASN with mobile topology	It improves the end-to-end success delivery rate of packet forwarding.	More powerful nodes are required

Table 1. Comparisons of major routing strategies.

2.2.1. *Geo-based routing*

Geo-based routing uses the position information of sensor nodes to find the best route from a source node to a destination node. Each node has to be aware of the target area, its own position, and all neighbors' positions. A node forwards packets to the target area in accordance with a certain policy based on the location information. At present, two common ways to obtain position information are using signal strength to estimate the relative coordinates and GPS. However, GPS cannot work well in underwater environments, and the relative coordinate of a node estimated by signal strength suffers from signal attenuation and noise interference.

2.2.2. *Source routing*

Source routing allows the sender of a packet to specify the route that the packet takes in the network partially or completely. The transmission path in is determined by the source node, and the identification of all relay nodes in the path is embedded in the packets. Relay nodes can forward packets according to the planned path instead of selecting next hop by themselves [1]. During the route construction phase, a source node establishes a route by flooding route request packets (RREQs) in the network. Upon receiving an RREQ, the destination node responds by sending a route reply packet (RREP), which carries the route attached in the RREQ back to the source node. It supports asymmetric channel and reduces the cost of route maintenance. However, it increases the packet overhead and the routing cost. In addition, the performance degrades rapidly with the increase of mobility.

2.2.3. *Hop-by-hop routing*

Hop-by-hop routing allows each relay node to select its next hop by itself. The selection of next hop is usually based on its local view of the network. It can make the network more scalable and flexible, but the final path may not be optimal.

2.2.4. *Clustered routing*

Clustered routing divides the sensor nodes into geographical or logical groups, and the selection of cluster head considers the position distribution of nodes and residual energy. In each group, a cluster head is used as a gateway to communicate with other groups, and sensor nodes in each group only need to transmit packets to the cluster head. This strategy has a good adaptability and reduces data redundancy.

2.2.5. *Cross-layer routing*

Although the hierarchical protocol stack structure is clear, scalable, robust, and easy to implement, the inter-layer information is difficult to be shared and not conducive to the global optimization of network performance. Cross-layer routing strategies take into account the functions and information available on other layers. Through the power control and frame collision control, it can achieve the relevant hierarchical interaction and minimize energy costs while maximizing the performance of the entire network [7].

2.2.6. *Reinforcement learning-based routing*

Reinforcement learning-based routing aims to find the most suitable route by using Q-learning algorithm to learn the network states and adapts to topology changes at runtime intelligently without any off-line training. During the routing process, a node analyzes its residual energy and energy distribution of a group of nodes, establishes a reinforcement function, and selects the appropriate node to forward packets [8]. This strategy can extend the lifetime of the network as much as possible.

2.2.7. *Opportunistic routing*

Opportunistic routing takes advantages of the packet transmission opportunities due to mobility and broadcast of nodes, not only determining the next hop but also selecting suitable neighbor nodes as a set of candidate forwarding nodes based on certain routing metrics. All nodes in the set receive packets and then the node with the highest priority completes the following transmission. It makes full use of channel broadcast characteristics to improve throughput and transmission reliability [9].

3. Single-sink–based UASNs

3.1. Static sink

A multi-path grid-based geographic routing protocol (MGGR) is proposed in Ref. [10]. It assumes that there is a sink fixed at the top in the middle of the grid, and the other sensor nodes are mobile and equipped with locator. Routing is performed in a grid-by-grid manner through gateways that use disjoint paths to relay data packets [10]. The disjoint paths and gateway election algorithm adopted in the protocol are helpful for balancing energy while achieving high delivery ratio and small end-to-end delay.

In Ref. [11], a location-based adaptive routing protocol (CARP) selects different paths for different levels of data packets, relying on the dynamic characteristics of underwater environments. CARP can adapt dynamic underwater environments, improve the quality of the network communication, and have better performance in transmission delay, energy consumption, and packet acceptance rate.

The hop-by-hop vector-based forwarding routing protocol (HH-VBF) is homogeneous to VBF which uses the notion of a "routing vector" to act as the axis of the "virtual pipe" [12]. However, it constructs different virtual pipes for the per-hop vector from each individual forwarder to the sink, instead of a single virtual pipe from the source to the sink as VBF. HH-VBF is less sensitive to the routing pipe radius threshold and has much better performance such as energy consumption and successful delivery rate than VBF in sparse networks.

A solution for routing joint control and node replacement decisions is proposed in Ref. [13] to minimize the node replacement costs and develop effective methods for practical implementation. Sensor nodes in the network are laid on the ocean floor to collectively gather and transmit seismic information. As the results indicate, it provides a lower average node replacement cost and meets higher service-level requirements, while it has a higher degree of simplicity.

A fault tolerant routing protocol (FTR) [14] assumes that the topology of the network is static and only a small fraction of the nodes is involved in routing. It detects substitutive links to revise the main path and build backup path. Simulations show that FTR achieves higher packet delivery ratio, small end-to-end delay, higher network throughput, and lower energy consumption than VBF and epidemic routing (ER) protocols.

3.2. Mobile sink

In Ref. [15], the authors proposed a mobicast routing protocol. All sensor nodes are randomly distributed in a 3-D environment, and an AUV travels along a user-defined route as a mobile sink to collect data from sensor nodes within a 3-D zone of relevance (ZOR) and wakes up sensors in the next 3-D ZOR [15]. In addition, only sensor nodes in the 3-D ZOR can be notified to enter the active mode to transmit collected data. For this reason, it has better performance such as a successful delivery rate, power consumption, and message overhead.

3.3. Geo-based routing

In Ref. [16], the authors proposed an integration method between localization and routing by using an AUV as an anchor node and "iterative localization". Each sensor node establishes multiple routes to the sink by extending control packets exchanged for localization and creates routing table to the sink in parallel with the localization process [16]. This method provides robustness to communication failure and decreases energy consumption between each localization.

A reliable and energy efficient vector-based forwarding protocol (REE-VBF) was proposed in Ref. [17]. The transmission way of data packets is local flooding, and an optimal node with the smallest desire coefficient is selected by establishing a set of adaptive selection mechanisms [17]. The optimal node forwards the packet at first to prohibit others transmitting repeatedly. Therefore, REE-VBF has a better performance on energy efficiency and reliability than others. In addition, it is suitable for the communication in shallow water areas.

In the directional flooding-based routing protocol (DFR) [18], a node knows its own location, its one-hop neighbors' location, and the location of a sink. It relies on a packet flooding scheme, with which the flooding zone is determined dynamically by the link quality among neighbors to increase the successful delivery ratio instead of relying on a path establishment between the source and sink node. The results show that DFR has a higher packet delivery ratio, less communication overhead, and shorter end-to-end delay than VBF.

A geographical optimized reflection-enabled routing protocol immune to link ambiguity (GORRILA) is proposed in Ref. [19], aiming to establish the best stable route from a source to a destination. It utilizes directional antennas to incorporate surface-reflected non-line-of-sight (NLOS) links in the routing process, instead of relying on the LOS link between one-hop neighbors, which adds delay to the data delivery time in establishing routes. In addition, a physical and medium access cross-layer scheme was also designed to optimize the route selection for maximum network throughput.

A sector-based routing with destination location prediction (SBR-DLP) protocol [20] was designed for a fully mobile topology network, in which each node is mobile and the destination node moves along a pre-planned route. In the routing process, the area around the current forwarder is divided into a plurality of sectors, wherein the first sector is halved by the vector from the transponder to the destination vector and the other sectors are tagged according to their angular differences from that vector. The SBR-DLP is highly adaptive to network dynamics and can improve the packet delivery ratio.

3.4. Source routing

A reliable and energy-efficient routing protocol (RER) was proposed in Ref. [21]. In the route discovery phase, a simple next-hop selection is introduced to select a node with the minimum packet transmission delay to the current node as its next-hop node. After a route is set up, multiple nodes in the network can also send data packets to the sink through the same established route instead of building a new route. As a result, it allows the real-time data to arrive at the destination earlier and extends the network lifetime. The simulations show that RER performs better than the traditional routing protocols in terms of packet delivery ratio, average end-to-end delay, and average energy consumption.

Another source routing protocol called local area source routing (LASR) [22] is proposed for a small area of UASN. In this case, a source node does not need to care about the outside area of its transmission range. For this reason, each node just needs to share information of its neighboring nodes and finds a minimum cost path up to a determined sub-destination within its transmission range. Simulation results show that the proposed solution can reduce energy consumption, in comparison with other 3-D on-demand routing protocols.

3.5. Hop-by-hop routing

In channel-aware depth-adaptive routing protocol (CDRP) [23], the source node constructs a virtual ideal path to the sink node when it has data to send. Besides, the noise and the speed of sound under different water depths are taken into account in the selection of forwarders to reduce end-to-end delay and improve packet delivery ratio.

In view of the impact of the beam width and 3-D direction of underwater sensors in UASNs, beam width and direction concerned routing (BDCR) proposed in Ref. [24] can achieve relatively high packet delivery rate and ensure reasonable energy consumption. On the selection of forwarder, the sensor node makes its preliminary decision not only on its own distance to the sink node but also on the distance of the previous sender to the sink node. Then, a correction mechanism based on beam width and direction is designed to make the final decision.

Chen et al. proposed a per-hop-based routing protocol called depth adaptive routing protocol (DARP) in Ref. [25], which takes not only the sound speed in different water depths but also the depth and the distance to the sink into consideration. In the routing process, a source node broadcasts the packets and then the neighbors wait for some time to decide whether they are eligible to forward the packets or not. The simulation results verify that DARP outperforms other routing protocols in terms of end-to-end delay.

UASNs-MIMO (UMIMO)-routing [26] utilizes multiple-input multiple-output-orthogonal frequency division multiplexing (MIMO-OFDM) links to adaptively leverage the trade-off between multiplexing and diversity gain. With cross-layer design, it adapts its behavior to the noise and interference in underwater channels to choose a suitable transmission mode and allocate transmit power on subcarriers [26]. Moreover, the energy consumption is minimized according to the cooperation of transmitter and receiver to achieve the desired QoS (quality of service) according to application needs and channel condition.

SEANAR [27] is an energy-efficient and topology-aware routing protocol. It assigns larger weights to nodes with higher connectivity to the sink and adopts a simple yet effective greedy approach for making routing decisions. Although this is a simple greedy approach that uses the degree information (i.e., the number of neighbors) as the criteria to choose the successor, simulation results show that SEANAR achieves higher packet delivery ratio and a lower energy consumption in comparison with greedy forwarding and VBF-based routing protocol.

Redundancy-based adaptive routing (RBAR) [28] is a routing protocol designed for underwater delay tolerant networks (DTNs). It adopts a binary tree-based forwarding procedure for the packet replication process, which allows a node to hold a packet as long as possible until it has to make another copy to satisfy its delay requirements [28]. Simulations show that RBAR can meet different delay requirements and achieve a good balance between delivery ratio, delay, and energy consumption.

3.6. Clustered network

A location-based clustering algorithm for data gathering (LCAD) [29] is a cluster-based routing protocol. The whole network is divided into 3-D grids and the sensor nodes are fixed at different depths. The sensor nodes at each tier are organized in clusters with multiple cluster heads. Data gathered from the sensor nodes are sent to their respective cluster heads and then cluster heads deliver the data to the sink via AUVs. Its performance depends on the positioning of the cluster heads. The simulation shows that it improves the network lifetime by at least five times as against a scenario, which does involve clustering.

A distributed underwater clustering scheme (DUCS) without GPS support was proposed in Ref. [30]. The nodes organize themselves into local clusters. A non-cluster head node forwards the packets to cluster heads in a single hop and then the cluster head transmits the packets to a sink via the relay of other cluster heads in a multi-hop mode. In order to solve a battery draining problem for the cluster head, the routing protocol performs random rotation of the cluster head. The simulation shows that DUCS can achieve a very high packet delivery ratio and minimize the proactive routing message exchange.

An energy optimized path unaware layered routing protocol (E-PULRP) [31] puts the sensor nodes into different layers in the form of concentric shells around a sink node. In each layer, an intermediate relay node is selected to deliver packets from the source node to the sink node. A mathematical framework is developed for energy consumption optimization. In comparison with other routing protocols for UASNs, E-PULRP is simpler and more topology independent.

The location unaware multi-hop routing protocol based on a hybrid, energy-efficient, distributed clustering approach (LUM-HEED) [32] is a new, homogeneous, multi-hop routing protocol. It can be adaptive to a hierarchal structure network model in which each node is initialized with a certain degree according to its distance to the sink. The difference between LUM-HEED and HEED protocols is that the nodes in HEED protocol must be location aware for communicating with the sink node. Note that the sensor nodes nearer to the sink have a higher degree. Simulation results show that LUM-HEED has better performance than HEED in terms of network lifetime and network traffic.

A mobility aware routing protocol, called temporary cluster-based routing (TCBR) [33], divides the sensor nodes in different clusters according to their locations. Ordinary nodes collect and forward the packets to a nearer cluster head and then courier nodes can move vertically and deliver them to the multiple sinks deployed on the water surface with a piston module. TCBR takes the advantage of multiple-sink architecture with requiring any location information of sensor nodes. This makes it suitable not only for stationary and mobile networks but also for the hybrid networks [33].

A staggered time division multiple access (TDMA) underwater medium access control (MAC) pro tocol with routing (STUMP-WR) [34] is a distributed and channel-scheduling routing protocol and designed for heavily loaded underwater networks. The sensor nodes select and schedule links to overlap communications by using a distributed algorithm for leveraging the long propagation delays [34]. The simulation shows that STUMP-WR outperforms several protocols proposed for underwater networks in terms of bits delivered per unit of energy and throughput.

4. Multi-sink–based UWSNs

4.1. Static UASNs

The depth-controlled routing protocol (DCR) in Ref. [35] uses the distance of the nodes to its nearest sink in a greedy forwarding strategy. In the network initialization phase, a sensor node moves to get a good and stable topology to improve the connectivity and routing performance. DCR shows better results than the current depth-based routing protocol (DBR) in terms of data delivery ratio, delay, and average number of redundant packets with the same energy per packet consumption.

Focused on energy balancing, the dual-sink efficient and balanced energy consumption technique (DSEBET) in Ref. [36] first establishes links between nodes based on their optimum distance value and then picked relay nodes based on their minimum distance value for the transmission of data. Long-distance nodes from one sink will share their data to another sink if they come in range of sink; otherwise, they will establish a multi-hop path for transmission of data to the respective sink [36]. In this way, energy is balanced to improve network lifetime and throughput.

A new routing protocol called MobiSink was proposed in Ref. [37] to tackle the problem of high energy consumption and reduce the instability period based on the deficiencies of depth-based routing and energy-efficient depth-based routing. Each of the four mobile sinks moves in its own region, following linear motion in the horizontal region to collect data from the nodes [37]. If the sink enters the transmission range of these nodes, the packet is forwarded directly from the high-level node to the sink. In this way, the data forwarding load is reduced on the middle node.

4.2. Mobile UASNs

In the dual-sink (DS) VBF protocol [38], all nodes are dynamic, and dual-sink architecture is deployed to increase the number of nodes that are participating in the data-forwarding procedure. In comparison with VBF, in DS-VBF, each node calculates its distance from the nearest sink and transmits the packets to it. It considers both residual energy and location information instead of just location information to discover an optimized path to save energy [38]. Thus, it can improve the packet delivery ratio, reduce end-to-end delay, and balance total energy consumption to prolong network lifetime.

The hop–by-hop dynamic addressing-based (H2-DAB) routing protocol in Ref. [2] chooses the next hop node in a greedy strategy for broadcasting based on the hop count, called HopID, from source nodes to sink nodes on the surface water. H2-DAB can easily handle the node movements and support multiple sink architecture. However, the problem of multi-hop routing still exists with which nodes near the sink drain more energy because they are used frequently.

4.3. Geo-based routing

Similar to the geo-based routing in single-sink–based UASNs, the geo-based routing in multi-sink UASNs uses the position information too. The difference is that in multi-sink–based UASNs, packets are sent to one of the sink nodes [1].

In Ref. [35], the DCR is a centralized and distributed geographic routing protocol, with depth adjustment-based topology control for recovery of invalid communication area. A greedy forwarding strategy is proposed for geographic routing.

The DS-VBF in Ref. [38] is a dynamic and geographical routing protocol, considering both residual energy and location information as priority factors to discover an optimized routing path to save energy. Every sensor node is aware of its own location, and each data packet contains the location information of the source, forwarder, and destination nodes. In addition, a range of fields used for node mobility notion is also known by each sensor node. Based on the simulation results in comparison with VBF, average end-to-end delay is reduced but remaining energy and packet reception ratio are increased.

5. Zero-sink–based UWSNs

A trajectory-aware communication scheme based on statistical inference to model position uncertainty, combined with a practical cross-layer optimization for a WHOI acoustic micro-modem, is adopted in a paradigm-changing geographic routing protocol [39] to minimize energy consumption. Acoustic communications are used to transfer information between gliders and finally to a surface station. Implemented and tested in the proposed underwater communication test bed, it shows improvement over other routing protocols with only statistical approach or cross-layer approach in terms of end-to-end reliability, throughput, and energy consumption [39].

An AUV-aided routing method integrated path planning (AA-RP) protocol [40] uses AUVs to collect data from sensor nodes following a dynamic path, which is planned by AUVs. It utilizes the cooperation of multi-tasks to reduce energy consumption and avoid hot spot and zone problem with a dynamic gateway node scheme. The simulation shows that, although the AA-RP does not require location information in the routing process, it can balance energy consumption, avoid hot point and hot zone problem, and save energy by combining multi-tasks with a good delivery ratio.

6. Cross-layer routing

The energy-efficient interference-aware routing protocol in Ref. [41] is a centralized cross-layer heuristic solution for an efficient use of the scarce resources of UASNs. It provides a class of scheduling, power control, and routing policies and selects the next transmission node by considering different delays in packet delivery, maximal node buffer size, distance to the sink, and channel usage. Such routing can increase the overall network throughput and outperforms others in terms of energy consumption and throughput.

A new geographical and distributed routing algorithm was tailored for the characteristics of 3-D underwater environment in Ref. [42]. A model characterizing the acoustic channel utilization efficiency allows setting the optimal packet size for underwater communication. Moreover, the problem of data gathering was investigated at the network layer by considering the cross-layer interactions with MAC layers, forward error correction schemes between the routing functions, and the characteristics of underwater acoustic channels [42]. In the light of different application requirements, two distributed routing algorithms were introduced to minimize energy consumption.

The gossiping in underwater acoustic mobile ad-hoc networks (GUWMANET) scheme in Ref. [43] realizes medium access and routing functionalities in a cross-layer design. It is based on impulse communication as a physical layer method. GUWMANET needs only 10 bits of additional overhead in combination with the generic underwater application language (GUWAL), which has a 16-bit header with a multi-cast source and destination address.

A multi-path power-control transmission (MPT) scheme [44] smartly combines power control with multi-path routing and packets at the destination. MPT is divided into three parts: multi-path routing, source-initiated power-control transmission, and destination packet combining. With carefully designed power-control strategies, MPT consumes lesser energy than the conventional one-path transmission scheme without retransmission [44]. Besides, since no hop-by-hop retransmission is allowed, MPT introduces much shorter delays than the traditional one-path scheme with retransmission.

The CARP in Ref. [45] is a distributed cross-layer solution for multi-hop delivery of data to a sink in underwater networks. Next-hop selection takes explicitly into account the history of data packet delivery, the link quality, and how successful a neighbor has been in forwarding data toward the sink [45]. The results show that CARP can achieve throughput efficiency that is up to twice the throughput of focused beam routing (FBR) and almost three times that of DBR. It also obtains remarkable performance improvements over FBR and DBR in terms of end-to-end packet latency and energy consumption.

7. Reinforcement learning-based routing

A novel Q-learning-based delay tolerant routing (QDTR) protocol [8] with predictions empowered by adaptive filters is adaptive and energy efficient. The adaptive filters are used to predict future neighbor contact. With the Q-learning agent, the routing protocol can adapt to changes in the network. Since the routing problem is formulated as a Markov decision process (MDP), in which the state space is composed by all the nodes in the network, QDTR is fully distributed without any central control. The simulation results have shown that QDTR yields significantly better network performance in energy consumption, end-to-end delay, and delivery ratio in comparison with most of the existing DTN routing protocols.

A multi-level routing protocol for acoustic-optical hybrid underwater wireless sensor networks (MURAO) [46] is a multi-level Q-learning-based routing protocol for a novel acoustic-optical hybrid UASN. The network is physically partitioned into several groups and logically divided into two layers. Taking advantage of the long range but slow acoustic transmission and fast optical communications with multi-level Q-learning, MURAO performs better than the flat Q-learning-based routing.

In Ref. [47], a Q-learning-based tracking scheme based on the buffer size and residual energy of the individual node was used to find the next forwarder. It aims to reduce the dropping on the packets, the number of forwarders, and energy consumption of the sensor nodes. The lifetime of the network is expected to increase.

Another Q-learning-based energy-efficient and lifetime-aware routing protocol (QELAR) was proposed in Ref. [48] to prolong the lifetime of networks. The residual energy of each sensor node as well as the energy distribution among a group of nodes is factored in the throughput routing process to calculate the reward function, which aids in selecting the adequate forwarders for packets. Compared with VBF, QELAR has a longer lifetime.

8. Opportunistic routing

The geographic and opportunistic routing with depth adjustment-based topology control for communication recovery over void regions (GEDAR) in Ref. [49] adjusts the topology by moving void nodes to new depths and using greedy opportunistic forwarding mechanisms to transmit packets. The communication void region occurs whenever the data is transferred to a node that is not closer to the destination than the node; the node located in a communication void region is called void node. Compared with the baseline routing protocols, GEDAR outperforms in data packet delivery ratio.

The void-aware pressure routing protocol (VAPR) in Ref. [50] exploits periodic beaconing to build directional trails toward the closest sonobuoy and features greedy opportunistic directional forwarding mechanisms for packet delivery. It can be efficiently performed even in the presence of voids [50]. The simulations show that VAPR outperforms existing schemes by significantly lowering the frequency of recovery fallbacks and effectively handling node mobility.

With the opportunistic-based DARP in Ref. [51], forwarding node selection is dynamic and independent for each node. DARP takes different acoustic signal speed, depth, and distance to sink into account to find the minimum end-to-end delay path, which may not be the shortest path directly from the source to the sink. Furthermore, it does not need to continuously maintain neighbors' information or to exchange control packets.

The HydroCast in Ref. [52] is a hydraulic pressure-based anycast routing protocol, which exploits measured pressure levels to route a packet upward to lower depths. The opportunistic routing mechanism can limit co-channel interference by selecting the subset of forwarder. The dead-end recovery method can guarantee the delivery. Because HydroCast uses adaptive timer setting at each hop, it is mainly used for depth-based communication with sparse network and performs better than DBR in terms of delivery ratio and end-to-end delay.

9. Conclusion and open issues

We summarize the routing strategies mentioned above in **Table 2**. Most routing strategies are suitable for a static UASN, and just a small scale of them can be applied in a mobile UASN. However, one of the most important characteristics of UASNs is mobility. From the adaptability of the application scenarios, geo-based routing protocols cannot work well in mobile UASNs due to frequent localization. In addition, the characteristics of acoustic channels are also the limitation of the design of geo-based routing protocols. Clustered routing protocols also do not perform well in mobile UASNs because of the grouping cost.

Cross-layer routing protocols may have good performance in static UASNs. Source-routing protocols are based on the location of sensor nodes and usually adapted to single-sink–based UASNs. On the opposite, hop-by-hop routing protocols can be applied in both single-sink and multi-sink–based UASNs.

Routing strategy\	Scenarios				
	Single-sink–based UASNs		Multi-sink–based UASNs		Zero-sink–based UASNs
	Static sink	Mobile sink	Static topology	Mobile topology	
Geo-based routing	✓		✓		✓
Source routing	✓				
Hop-by-hop routing	✓	✓	✓	✓	
Clustered network	✓		✓		
Cross-layer routing	✓				✓
Reinforcement learning-based routing	✓				
Opportunistic routing				✓	

Table 2. Categories of routing strategies.

Reinforcement learning-based routing protocols have good adaptability. It may be applied to dynamic networks by the design of reward functions.

The above discussion shows that it is impossible to design just one or two routing protocols that can cost-effectively support all underwater application scenarios. Many routing proposals for UASNs are only in the simulation phase and have not been testified in the actual environments. Researchers still follow the design philosophy of routing protocols for terrestrial wireless networks, which is not enough for UASNs. An optimal design must take into account each different favorable feature available in different scenarios.

Acknowledgements

This work is supported (in part) by National Natural Science Foundation of China (NSFC) under Grant 61472237.

Author details

Qian Lu, Feng Liu, Ying Zhang and Shengming Jiang*

*Address all correspondence to: smjiang@shmtu.edu.cn

College of Information Engineering, Shanghai Maritime University, Shanghai, China

References

[1] Qian L, Jiang SM. A review of routing protocols of underwater acoustic sensor networks from application perspective. In: 2016 IEEE International Conference on Communication Systems (ICCS); 14-16 December 2016; Shenzhen, China: IEEE; 2017. pp. 1-6. DOI: 10.1109/ICCS.2016. 7833633

[2] Ayaz M, Abdullah A. Hop-by-hop dynamic addressing based (H2-DAB) routing protocol for underwater wireless sensor networks. In: 2009 International Conference on Information and Multimedia Technology; 16-18 December 2009; Jeju Island, South Korea: IEEE; 2010. pp. 436-441. DOI: 10.1109/ICIMT.2009.70

[3] Tonghong L. Multi-sink opportunistic routing protocol for underwater mesh network. In: 2008 International Conference on Communications, Circuits and Systems; 25-27 May 2008; Fujian, China: IEEE; 2008. pp. 405-409. DOI: 10.1109/ICCCAS.2008.4657802

[4] Li N, Martínez J-F, Meneses Chaus J, Eckert M. A Survey on underwater acoustic sensor network routing protocols. Sensors. Mar. 2016;**16**(3):414. DOI: 10.3390/s16030414

[5] Akyildiz IF, Pompili D, Melodia T. State-of-the-art in protocol research for underwater acoustic sensor networks. In: Proceedings of the 1st ACM International Workshop on Underwater Networks; Los Angeles, CA, USA. New York, NY, USA: ACM; 2006. pp. 7-16. DOI: 10.1145/1161039.1161043

[6] Muhammad A, Imran B, Azween A, Ibrahima F. A survey on routing techniques in underwater wireless sensor networks. Journal of Network and Computer Applications. 2011;**34**(6):1908-1927. DOI: http://dx.doi.org/10.1016/j.jnca.2011.06.009

[7] Wang G, Liu K. Routing protocol based on cross-layer design in wireless sensor networks. Journal of Beijing University of Aeronautics and Astronautics. 2010;**36**(6):732-735. DOI: 10.13700/j.bh.1001-5965.2010.06.027. [in Chinese]

[8] Tiansi H, Yunsi F. An adaptive routing protocol based on connectivity prediction for underwater disruption tolerant networks. In: 2013 IEEE Global Communications Conference (GLOBECOM); 9-13 December 2013; Atlanta, GA, USA: IEEE; 2014. pp. 65-71. DOI: 10.1109/GLOCOM.2013.6831049

[9] Tian K, Zhang B, Ma J, Yao Z. Opportunistic routing protocols for wireless multihop networks. Journal of Software. 2010;**21**(10):2542-2553. DOI: 10.3724/SP.J.1001.2010.03740. [in Chinese]

[10] Al-Salti F, Alzeidi N, Arafeh B. A new multipath grid-based geographic routing protocol for underwater wireless sensor networks. In: 2014 International Conference on Cyber- Enabled Distributed Computing and Knowledge Discovery; 13-15 October 2014; Shanghai, China: IEEE; 2014. pp. 331-336. DOI: 10.1109/CyberC.2014.64

[11] Li J, Du X, Xing J, Yang Q. Location based adaptive routing protocol for underwater acoustic sensor networks. In: International Conference on Automatic Control and Artificial Intelligence (ACAI 2012); 3-5 March 2012; Xiamen, China: IET; 2013. pp. 1315-1319. DOI: 10.1049/cp.2012.1221

[12] Nicolaou N, See A, Xie P, Cui JH, Maggiorini D. Improving the robustness of location based routing for underwater sensor networks. In: OCEANS 2007—Europe; 18-21 June 2007; Aberdeen, UK: IEEE; 2007. pp. 1-6. DOI: 10.1109/OCEANSE.2007.4302470

[13] Mohapatra AK, Gautam N, Gibson RL. Combined routing and node replacement in energy-efficient underwater sensor networks for seismic monitoring. IEEE Journal of Oceanic Engineering. 2013;**38**(1):80-90. DOI: 10.1109/JOE.2012.2208850

[14] Xu M, Liu G. Fault tolerant routing in three-dimensional underwater acoustic sensor networks. In: 2011 International Conference on Wireless Communications and Signal Processing (WCSP); 9-11 November 2011; 2011 International Conference on Wireless Communications and Signal Processing (WCSP). IEEE; 2011. pp. 1-5. DOI: 10.1109/WCSP.2011.6096859

[15] Chen YS, Lin YW. Mobicast routing protocol for underwater sensor networks. IEEE Sensors Journal. 2013;**13**(2):737-749. DOI: 10.1109/JSEN.2012.2226877

[16] Hirai S, Tanigawa Y, Tode H. Integration method between localization and routing in underwater sensor network. In: 2012 IEEE 15th International Conference on Computational Science and Engineering; 5-7 December 2012; Nicosia, Cyprus: IEEE; 2013. pp. 689-693. DOI: 10.1109/ICCSE.2012.99

[17] Wang P, Fu DH, Zhao CQ, Xing JC, Yang QL, Du XF. A reliable and efficient routing protocol for underwater acoustic sensor networks. In: 2013 IEEE International Conference on Cyber Technology in Automation, Control and Intelligent Systems; 26-29 May 2013; Nanjing, China: IEEE; 2014. pp. 185-190. DOI: 10.1109/CYBER.2013.6705443

[18] Daeyoup H, Dongkyun K. DFR: Directional flooding-based routing protocol for underwater sensor networks. In: OCEANS 2008; 15-18 September 2008; Quebec City, QC, Canada: IEEE; 2009. pp. 1-7. DOI: 10.1109/OCEANS.2008.5151939

[19] Emokpae L, Younis M. Signal reflection-enabled geographical routing for underwater sensor networks. In: 2012 IEEE International Conference on Communications (ICC); 10-15 June 2012; Ottawa, ON, Canada: IEEE; 2012. pp. 147-151. DOI: 10.1109/ICC.2012.63637

[20] Chirdchoo N, Soh WS, Chua KC. Sector-based routing with destination location prediction for underwater mobile networks. In: 2009 International Conference on Advanced Information Networking and Applications Workshops; 26-29 May 2009; Bradford, UK: IEEE; 2009. pp. 1148-1153. DOI: 10.1109/WAINA.2009.105

[21] Wei Y, Kim DS. Reliable and energy-efficient routing protocol for underwater acoustic sensor networks. In: 2014 International Conference on Information and Communication Technology Convergence (ICTC); 22-24 October 2014; Busan, South Korea: IEEE; 2014. pp. 738-743. DOI: 10.1109/ICTC.2014. 6983274

[22] Eunjo K, Jonggu K, Poh KC, Seongeun Y, Daeyoung K. Energy efficient local area source routing protocol of underwater sensor networks in the deep ocean. In: 2007 International Symposium on Communications and Information Technologies; 17-19 October 2007; Sydney, NSW, Australia: IEEE; 2007. pp. 948-953. DOI: 10.1109/ISCIT.2007.4392152

[23] Chen YD, Chen YW, Lien CY, Shih KP. A channel-aware depth-adaptive routing protocol for underwater acoustic sensor networks. In: OCEANS 2014—TAIPEI; 7-10 April 2014; Taipei, Taiwan: IEEE; 2014. pp. 1-6. DOI: 10.1109/OCEANS-TAIPEI.2014.6964460

[24] Zhang S, Li D. A beam width and direction concerned routing for underwater acoustic sensor networks. In: 2013 IEEE 9th International Conference on Mobile Ad-hoc and Sensor Networks; 11-13 December 2013; Dalian, China: IEEE; 2014. pp. 17-24. DOI: 10.1109/MSN.2013.21

[25] Chen YD, Lien CY, Wang CH, Shih KP. DARP: A depth adaptive routing protocol for large-scale underwater acoustic sensor networks. In: 2012 Oceans—Yeosu; 21-24 May 2012; Yeosu, South Korea: IEEE; 2012. pp. 1-6. DOI: 10.1109/OCEANS-Yeosu.2012.6263505

[26] Kuo LC, Melodia T. Cross-layer routing on MIMO-OFDM underwater acoustic links. In: 2012 9th Annual IEEE Communications Society Conference on Sensor, Mesh and Ad

Hoc Communications and Networks (SECON); 18-21 June 2012; Seoul, South Korea: IEEE; 2012. pp. 227-235. DOI: 10.1109/SECON.2012.6275782

[27] Wu X, Chen G, Chen J. Energy-efficient and topology-aware routing for underwater sensor networks. In: 2010 Proceedings of 19th International Conference on Computer Communications and Networks; 2-5 August 2010; Zurich, Switzerland, Switzerland: IEEE; 2010. pp. 1-6. DOI: 10.1109/ICCCN.2010.5560016

[28] Zheng G, Zheng P, Bing W, Cui JH, Wu J. Adaptive routing in underwater delay tolerant sensor networks. In: 2011 6th International ICST Conference on Communications and Networking in China (CHINACOM); 17-19 August 2011; Harbin, China: IEEE; 2012. pp. 1044-1051. DOI: 10.1109/ChinaCom.2011.6158310

[29] Anupama KR, Sasidharan A, Vadlamani S. A location-based clustering algorithm for data gathering in 3D underwater wireless sensor networks. In: 2008 International Symposium on Telecommunications; 27-28 August 2008; Tehran, Iran: IEEE; 2008. pp. 343-348. DOI: 10.1109/ISTEL.2008.4651325

[30] Domingo MC, Prior R. Design and analysis of a GPS-free routing protocol for underwater wireless sensor networks in deep water. In: 2007 International Conference on Sensor Technologies and Applications (SENSORCOMM 2007); 14-20 October 2007; Valencia, Spain: IEEE; 2007. pp. 215-220. DOI: 10.1109/SENSORCOMM.2007.4394923

[31] Gopi S, Govindan K, Chander D, Desai UB, Merchant SN. E-PULRP: energy optimized path unaware layered routing protocol for underwater sensor networks. IEEE Transactions on Wireless Communications. 2010;**9**(11):3391-3401. DOI: 10.1109/TWC.2010.091510.090452

[32] Ce W, Gongliang L. LUM-HEED: A location unaware, multi-hop routing protocol for underwater acoustic sensor networks. In: Proceedings of 2011 International Conference on Computer Science and Network Technology; 24-26 December 2011; Harbin, China: IEEE; 2012. pp. 2336-2340. DOI: 10.1109/ICCSNT.2011.6182441

[33] Ayaz M, Abdullah A, Jung LT. Temporary cluster based routing for underwater wireless sensor networks. In: 2010 International Symposium on Information Technology; 15-17 June 2010; Kuala Lumpur, Malaysia: IEEE; 2010. pp. 1009-1014. DOI: 10.1109/ITSIM.2010.5561598

[34] Kredo II K, Mohapatra P. Distributed scheduling and routing in underwater wireless networks. In: 2010 IEEE Global Telecommunications Conference GLOBECOM 2010; 6-10 December 2010; Miami, FL, USA: IEEE; 2011. pp. 1-6. DOI: 10.1109/GLOCOM.2010.5685230

[35] Coutinho RWL, Vieira LFM, Loureiro AAF. DCR: Depth-controlled routing protocol for underwater sensor networks. In: 2013 IEEE Symposium on Computers and Communications (ISCC); 7-10 July 2013; Split, Croatia: IEEE; 2014. pp. 453-458. DOI: 10.1109/ISCC.2013.6754988

[36] Khan MA, Javaid N, Majid A, Imran M, Alnuem M. Dual sink efficient balanced energy technique for underwater acoustic sensor networks. In: 2016 30th International

Conference on Advanced Information Networking and Applications Workshops (WAINA); 23-25 March 2016; Crans-Montana, Switzerland: IEEE; 2016. pp. 551-556. DOI: 10.1109/WAINA.2016.156

[37] Shah PM, Ullah I, Khan T, Hussain MS, Khan ZA, Qasim U, Javaid N. MobiSink: Cooperative routing protocol for underwater sensor networks with sink mobility. In: 2016 IEEE 30th International Conference on Advanced Information Networking and Applications (AINA); 23-25 March 2016; Crans-Montana, Switzerland: IEEE; 2016. pp. 189-197. DOI: 10.1109/AINA.2016.167

[38] Pouryazdanpanah KM, Anjomshoa M, Salehi SA, Afroozeh A, Moshfegh GM. DS-VBF: Dual sink vector-based routing protocol for underwater wireless sensor network. In: 2014 IEEE 5th Control and System Graduate Research Colloquium; 11-12 August 2014; Shah Alam, Malaysia: IEEE; 2014. pp. 227-232. DOI: 10.1109/ICSGRC.2014.6908727

[39] Chen B, Hickey PC, Pompili D. Trajectory-aware communication solution for underwater gliders using WHOI micro-modems. In: 2010 7th Annual IEEE Communications Society Conference on Sensor, Mesh and Ad Hoc Communications and Networks (SECON); 21-25 June 2010; Boston, MA, USA, USA: IEEE; 2010. pp. 1-9. DOI: 10.1109/SECON.2010.5508262

[40] Zhuo W, Hongmei G, Longjie J, Xiaoning F. AUV-aided communication method for underwater mobile sensor network. In: OCEANS 2016—Shanghai; 10-13 April 2016; Shanghai, China: IEEE; 2016. pp. 1-7. DOI: 10.1109/OCEANSAP.2016.7485345

[41] Shashaj A, Petroccia R, Petrioli C. Energy efficient interference-aware routing and scheduling in underwater sensor networks. In: 2014 Oceans—St. John's; 14-19 September 2014; St. John's, NL, Canada: IEEE; 2015. pp. 1-8. DOI: 10.1109/OCEANS.2014.7003110

[42] Pompili D, Melodia T, Akyildiz IF, Ian F.. Routing algorithms for delay-insensitive and delay-sensitive applications in underwater sensor networks. In: Proceedings of the 12th Annual International Conference on Mobile Computing and Networking; Los Angeles, CA, USA. New York, NY, USA: ACM; 2006. pp. 298-309. DOI: 10.1145/1161089.1161123

[43] Goetz M, Nissen I. GUWMANET-multicast routing in underwater acoustic networks. In: 2012 Military Communications and Information Systems Conference (MCC); 8-9 October 2012; Gdansk, Poland: IEEE; 2012. p. 1-8.

[44] Zhou Z, Peng Z, Cui JH, Shi Z. Efficient multipath communication for time-critical applications in underwater acoustic sensor networks. IEEE/ACM Transactions on Networking. 2011;**19**(1):28-41. DOI: 10.1109/TNET.2010.2055886

[45] Basagni S, Petrioli C, Petroccia R, Spaccini D. Channel-aware routing for underwater wireless networks. In: 2012 Oceans—Yeosu; 21-24 May 2012; Yeosu, South Korea: IEEE; 2012. pp. 1-9. DOI: 10.1109/OCEANS-Yeosu.2012.6263538

[46] Hu T, Fei Y. MURAO: A multi-level routing protocol for acoustic-optical hybrid underwater wireless sensor networks. In: 2012 9th Annual IEEE Communications Society Conference on Sensor, Mesh and Ad Hoc Communications and Networks

(SECON); 18-21 June 2012; Seoul, South Korea: IEEE; 2012. pp. 218-226. DOI: 10.1109/SECON.2012.6275781

[47] Thangarajan R, Siva T, Boopalachakaravarthy R. Adaptive energy efficient routing protocol with extended lifetime in underwater sensor networks. In: 2013 International Conference on Communication Systems and Network Technologies; 6-8 April 2013; Gwalior, India: IEEE; 2013. pp. 322-326. DOI: 10.1109/CSNT.2013.74

[48] Hu T, Fei Y. QELAR: A machine-learning-based adaptive routing protocol for energy-efficient and lifetime-extended underwater sensor networks. IEEE Transactions on Mobile Computing. 2010;**9**(6):796-809. DOI: 10.1109/TMC.2010.28

[49] Coutinho RWL, Boukerche A, Vieira LFM, Loureiro AAF. Geographic and opportunistic routing for underwater sensor networks. IEEE Transactions on Computers. 2016;**65**(2):548-561. DOI: 10.1109/TC.2015.2423677

[50] Noh Y, Lee U, Wang P, Choi BSC, Gerla M. VAPR: Void-aware pressure routing for underwater sensor networks. IEEE Transactions on Mobile Computing. 2013;**12**(5):895-908. DOI: 10.1109/TMC.2012.53

[51] Yang CH, Ssu KF, Lin YY. A delay-awareness routing protocol in intermittently connected underwater acoustic sensor networks. In: 2013 IEEE 19th Pacific Rim International Symposium on Dependable Computing; 2-4 December 2013; Vancouver, BC, Canada: IEEE; 2014. pp. 138-39. DOI: 10.1109/PRDC.2013.31

[52] Lee U, Wang P, Noh Y, Vieira LFM, Gerla M, Cui JH. Pressure routing for underwater sensor networks. In: 2010 Proceedings IEEE INFOCOM; 14-19 March 2010; San Diego, CA, USA: IEEE; 2010. pp. 1-9. DOI: 10.1109/INFCOM.2010.5461986

Chapter 3

Nonlinear Unknown-Input Observer-Based Systems for Secure Communication

Robert N.K. Loh and Manohar K. Das

Additional information is available at the end of the chapter

http://dx.doi.org/10.5772/intechopen.69239

Abstract

Secure communication employing chaotic systems is considered in this chapter. Chaos-based communication uses chaotic systems as its backbone for information transmission and extraction, and is a field of active research and development and rapid advances in the literature. The theory and methods of synchronizing chaotic systems employing unknown input observers (UIOs) are investigated. New and novel results are presented. The techniques developed can be applied to a wide class of chaotic systems. Applications to the estimation of a variety of information signals, such as speech signal, electrocardiogram, stock price data hidden in chaotic system dynamics, are presented.

Keywords: chaotic secure communication, underwater acoustic communication, chaos, unknown input observers, nonlinear observers, reduced-order observers

1. Introduction

With the advances in computing and communication technologies, among others, underwater acoustic communication (UAC) techniques [1–6] have emerged as the predominant mode of underwater communication because of its one key advantage over conventional electromagnetic communication, namely, relatively low attenuation of acoustic waves in water. However, their performance is severely affected by a number of factors, including limited channel bandwidth, time-varying channel characteristics, complex ambient noise, and multipath distortion that results from multiple reflections of sound waves from top and bottom surfaces of water, especially in a relatively shallow waterbody.

Over the past decade, chaos-based underwater acoustic communication (CUAC) techniques have attracted a lot of interest from a number of researchers [7–12], because such techniques are potentially more cost-effective (for example, requiring lesser number of component modules) compared with conventional communication schemes. The CUAC techniques proposed to date can be broadly divided into two categories, namely, coherent detection based CUAC methods [7], and non-coherent detection based CUAC techniques. The coherent detection based methods rely on synchronization to reconstruct a copy of the transmitted signal at the receiver end, whereas non-coherent detection methods [8–12] utilize a variety of data recovery methods without requiring any synchronized reconstruction of the transmitted message.

In this chapter, we focus our attention on the synchronization based CUAC techniques, especially on observer-based synchronization methods, because the underlying theory is very well understood and has proven to be reliable and robust in many control applications. Also, such methods may potentially turn out to be easier to implement, as compared with many non-coherent CUAC techniques.

At the outset, we should point out that the main goal of this chapter is to present the fundamental concepts of observer-based chaotic synchronization and their applications to secure chaotic communication. With this in mind and owing to space limitation, we omit discussion of the robustness issues [13–23] of such techniques here. However, we should point out that the theory of robust observer design in the presence of noise and uncertainties has been well researched in control literature, and these ideas are deemed to be useful for synchronized based CUAC as well [18–23].

The methodologies used for CUAC have many things in common with chaos-based wired and wireless communication. Research and development in these fields have been advancing rapidly in the literature [7–16]. In contrast to conventional communication systems which use sinusoidal carriers to transmit information, chaos-based communication uses chaotic systems as its backbone for information transmission and recovery. The advantages of employing chaos-based systems include, among others, (i) the communication is difficult to detect and decrypt; (ii) the transmission is hidden from unauthorized receivers; (iii) the communication is more resistant to jamming and interferences because of the broadband characteristics of the chaos-based carriers. The advantages above are due to the following characteristics: (i) a chaotic system is dissipative; (ii) chaotic systems have unstable equilibrium points; (iii) its trajectories are aperiodic and bounded; and (iv) its trajectories have a sensitive dependence on their initial conditions, i.e., trajectories originated from slightly different initial conditions will soon become totally different. We remark that some of these characteristics may, in fact, be undesirable.

The organization of this chapter is as follows. Section 2 introduces three nonlinear chaotic systems that are utilized for designing chaotic communication systems in follow-up sections. Next a general discussion of unknown input observers is presented in Section 3. Section 4 presents the theory and design of unknown input observers for chaotic secure communication. Finally, the conclusions and plan for future research are provided in Section 5.

2. Nonlinear systems with application in chaotic communication

Consider a general nonlinear system described by

$$\begin{cases} \dot{\mathbf{x}} - \mathbf{f}(\mathbf{x}, \mathbf{d}) \\ \mathbf{y} = \mathbf{h}(\mathbf{x}, \mathbf{d}), \end{cases} \tag{1}$$

where $\mathbf{x} \in \mathbb{R}^n$ is the system state vector, $\mathbf{y} \in \mathbb{R}^m$ the output measurement, $\mathbf{d} \in \mathbb{R}^r$ an unknown disturbance vector which can be treated as a message vector that carries useful information; $\mathbf{f} : \mathbb{R}^n \times \mathbb{R}^r \rightarrow \mathbb{R}^n$ is a smooth vector field, $\mathbf{h} : \mathbb{R}^n \times \mathbb{R}^r \rightarrow \mathbb{R}^m$ a smooth function, $\mathbf{f}(\mathbf{0}, \mathbf{0}) = \mathbf{0}$ and $\mathbf{h}(\mathbf{0}, \mathbf{0}) = \mathbf{0}$.

The unknown disturbance $\mathbf{d}$ in (1) is assumed to be generated by the exosystem

$$\begin{cases} \mathbf{d} = \mathbf{M}\mathbf{m}, \\ \dot{\mathbf{m}} = \mathbf{f}_m(\mathbf{m}, \mathbf{x}), \end{cases} \tag{2}$$

where $\mathbf{m} \in \mathbb{R}^r$ is the message state, $\mathbf{M} \in \mathbb{R}^{r \times r}$ is a "picking matrix" that picks the appropriate components m_i of $\mathbf{m}$ to form $\mathbf{d}$, $\mathbf{f}_m : \mathbb{R}^r \times \mathbb{R}^n \rightarrow \mathbb{R}^r$ is a smooth vector field, and $\mathbf{f}_m(\mathbf{0}, \mathbf{0}) = \mathbf{0}$.

Eqs. (1) and (2) is widely used for the design of linear and nonlinear observers, unknown input observers (UIO), and unknown input observers for secure communication [24–46]. When applied to the design of unknown input observers (UIOs) for secure communication based on chaotic systems, (1) and (2) can be combined and expressed as

$$\begin{cases} \dot{\mathbf{x}} = \mathbf{f}(\mathbf{x}, \mathbf{m}) = \mathbf{f}(\mathbf{x}) + \mathbf{B}_m(\mathbf{x})\mathbf{M}\mathbf{m} = \mathbf{A}\mathbf{x} + \mathbf{g}(\mathbf{x}) + \mathbf{B}_m(\mathbf{x})\mathbf{M}\mathbf{m}, \\ \dot{\mathbf{m}} = \mathbf{f}_m(\mathbf{m}, \mathbf{x}) = \mathbf{A}_m\mathbf{m} + \mathbf{\Psi}\mathbf{x}, \\ \mathbf{y} = \mathbf{h}(\mathbf{x}, \mathbf{m}), \quad (\mathbf{d} = \mathbf{M}\mathbf{m}), \end{cases} \tag{3}$$

where $\mathbf{Ax}$ is the linear part of $\mathbf{f(x)}$, $\mathbf{g}(\mathbf{x}) \in \mathbb{R}^{n \times 1}$ and $\mathbf{B}_m(\mathbf{x}) \in \mathbb{R}^{n \times r}$, while $\mathbf{m}$ ($\mathbf{d} = \mathbf{Mm}$) is now treated as the message signal, and $\mathbf{f}_m(\mathbf{m}, \mathbf{x}) = \mathbf{A}_m\mathbf{m} + \mathbf{\Psi}\mathbf{x}$, where $\mathbf{A}_m \in \mathbb{R}^{r \times r}$ is a constant matrix. The linear model in the second equation is commonly used in the literature, see for example [25]. In many applications, the message model can be simplified by setting $\mathbf{A}_m = \mathbf{0}$ and $\mathbf{\Psi} = \mathbf{0}$. Further, (3) may become a system with *state-dependent* or *multiplicative* and/or *additive* message signals depending on $\mathbf{B}_m(\mathbf{x})$. If $\mathbf{B}_m(\mathbf{x}) = \mathbf{B}_m$, where $\mathbf{B}_m$ is a constant matrix, then (3) is a system with only *additive* messages.

The following three chaotic systems in the form of (3) will be utilized for designing chaotic communication systems in this chapter.

(1) Rossler system [47]

The Rossler system described by

$$\dot{\mathbf{x}} = \mathbf{f}(\mathbf{x}) = \begin{bmatrix} -x_2 - x_3 \\ x_1 + ax_2 \\ -cx_3 + x_1x_3 + b \end{bmatrix} \tag{4}$$

can be modified by chaotic parameter modulation resulting in a system with *state-dependent (multiplicative)* and *additive* messages as

$$\begin{cases} \dot{\mathbf{x}} = \mathbf{f}(\mathbf{x},\mathbf{m}) = \begin{bmatrix} -x_2 - x_3 \\ x_1 + ax_2 \\ -(c-m_1)x_3 + x_1x_3 + (b+m_2) \end{bmatrix} = \begin{bmatrix} 0 & -1 & -1 \\ 1 & a & 0 \\ 0 & 0 & -c \end{bmatrix}\mathbf{x} + \begin{bmatrix} 0 \\ 0 \\ m_1x_3 + x_1x_3 + b \end{bmatrix} + \begin{bmatrix} 0 & 0 \\ 0 & 0 \\ 0 & 1 \end{bmatrix}\mathbf{m}, \\ \dot{\mathbf{m}} = \mathbf{A}_m\mathbf{m}, \\ \mathbf{y} = \mathbf{h}(\mathbf{x},\mathbf{m}), \end{cases} \tag{5}$$

where $\mathbf{\Psi} = \mathbf{0}$, and the chaotic parameters are given by $\{a,\ b,\ c\} = \{0.2,\ 0.2,\ 5.7\}$ [43] or $\{0.398,\ 2,\ 4\}$ [42]. Note that the Rossler system (4) contains only one nonlinear term. See also [48] for more details.

(2) Genesio-Tesi system [49]

The Genesio-Tesi system given by

$$\dot{\mathbf{x}} = \mathbf{f}(\mathbf{x}) = \begin{bmatrix} x_2 \\ x_3 \\ -cx_1 - bx_2 - ax_3 + x_1^2 \end{bmatrix} \tag{6}$$

can be modified in the form of (3) with state-dependent and additive message signals as,

$$\begin{cases} \dot{\mathbf{x}} = \mathbf{f}(\mathbf{x},\mathbf{m}) = \begin{bmatrix} x_2 \\ x_3 + m_1 \\ -cx_1 - (b-m_1)x_2 - ax_3 + x_1^2 + m_2 \end{bmatrix} = \mathbf{f}(\mathbf{x}) + \begin{bmatrix} 0 \\ 0 \\ -x_2 \end{bmatrix} m_1 + \begin{bmatrix} 0 & 0 \\ 1 & 0 \\ 0 & 1 \end{bmatrix}\begin{bmatrix} m_1 \\ m_2 \end{bmatrix}, \\ \dot{\mathbf{m}} = \mathbf{A}_\mathbf{m}\mathbf{m}, \\ \mathbf{y} = \mathbf{h}(\mathbf{x},\mathbf{m}), \end{cases} \tag{7}$$

where $\mathbf{\Psi} = \mathbf{0}$, a, b and c are the chaotic parameters satisfying $ab < c$ and are given by {a, b, c}={1.2, 2.92, 5.7} [49]. Note that, without the nonlinear term x_1^2, the Genesio-Tesi system (6) is a linear time-invariant (LTI) system and is a state-space realization of the transfer function $G(s) = 1/(s^3 + as^2 + bs + c)$.

(3) Chua circuit [50]

The Chua circuit

$$\dot{\mathbf{x}} = \mathbf{f}(\mathbf{x}) = \begin{bmatrix} \alpha(x_2 - x_1^3 - cx_1) \\ x_1 - x_2 + x_3 \\ -\beta x_2 \end{bmatrix} \tag{8}$$

may be modulated in a form with state-dependent and additive messages as

$$\begin{cases} \dot{\mathbf{x}} = \mathbf{f}(\mathbf{x}, \mathbf{m}) = \begin{bmatrix} \alpha(x_2 - x_1^3 - cx_1) + m_1 \\ x_1 - x_2 + x_3 \\ -(\beta + m_2)x_2 \end{bmatrix} = \mathbf{f}(\mathbf{x}) + \begin{bmatrix} 1 \\ 0 \\ 0 \end{bmatrix} m_1 + \begin{bmatrix} 0 \\ 0 \\ -x_2 \end{bmatrix} m_2, \\ \dot{m} - A_m m, \\ \mathbf{y} = \mathbf{h}(\mathbf{x}, m), \end{cases} \tag{9}$$

where $\alpha = 10$, $\beta = 16$ and $c = -0.14$ are the chaotic parameters. A different modification scheme is given in Ref. [51].

It is noted that, although chaotic systems are sensitive to variations of their chaotic parameters $\mathbf{p} = \{p_i\}$, most systems do accommodate suitable modifications of some of these parameters. This property has precisely been exploited for the designs of UIOs for secure communication and many control-based synchronization schemes in the literature.

3. General unknown-input observers (UIOs)

Consider (3), which can be expressed more compactly as,

$$\begin{cases} \dot{\mathbf{w}} = \mathbf{f}_w(\mathbf{w}), \\ \mathbf{y} = \mathbf{h}(\mathbf{w}), \end{cases} \tag{10}$$

where

$$\mathbf{w} = \begin{bmatrix} \mathbf{x} \\ \hline \mathbf{m} \end{bmatrix}, \ \mathbf{f}_w(\mathbf{w}) = \begin{bmatrix} \mathbf{f}(\mathbf{x}, \mathbf{m}) \\ \hline \mathbf{f}_m(\mathbf{m}, \mathbf{x}) \end{bmatrix}, \text{ and } \mathbf{h}(\mathbf{w}) = \mathbf{h}(\mathbf{x}, \mathbf{m}).$$

Consider a Luenberger-like nonlinear observer for (3) given by [27–31, 34],

$$\begin{cases} \begin{bmatrix} \dot{\hat{\mathbf{x}}} \\ \hline \dot{\hat{\mathbf{m}}} \end{bmatrix} = \begin{bmatrix} \mathbf{f}(\hat{\mathbf{x}}, \hat{\mathbf{m}}) \\ \hline \mathbf{f}_m(\hat{\mathbf{m}}, \hat{\mathbf{x}}) \end{bmatrix} + \begin{bmatrix} \mathbf{L}_{1o}(\cdot) \\ \hline \mathbf{L}_{2o}(\cdot) \end{bmatrix} [\mathbf{y} - \mathbf{h}(\hat{\mathbf{x}}, \hat{\mathbf{m}})], \\ \mathbf{y} = \mathbf{h}(\mathbf{x}, \mathbf{m}), \end{cases} \tag{11}$$

or more compactly as, with (10),

$$\begin{cases} \dot{\hat{\mathbf{w}}} = \mathbf{f}_w(\hat{\mathbf{w}}) + \mathbf{L}_o(\cdot)[\mathbf{y} - \mathbf{h}(\hat{\mathbf{w}})], \\ \mathbf{y} = \mathbf{h}(\mathbf{w}), \end{cases} \tag{12}$$

where $\hat{\mathbf{w}} = \begin{bmatrix} \hat{\mathbf{x}} \\ \hline \hat{\mathbf{m}} \end{bmatrix}$ is an estimate of $\mathbf{w} = \begin{bmatrix} \mathbf{x} \\ \hline \mathbf{m} \end{bmatrix}$, $\mathbf{f}_w(\hat{\mathbf{w}}) = \begin{bmatrix} \mathbf{f}(\hat{\mathbf{x}}, \hat{\mathbf{m}}) \\ \hline \mathbf{f}_m(\hat{\mathbf{m}}, \hat{\mathbf{x}}) \end{bmatrix}$, and $\mathbf{L}_o(\cdot) = \begin{bmatrix} \mathbf{L}_{1o}(\cdot) \\ \hline \mathbf{L}_{2o}(\cdot) \end{bmatrix}$ is the observer gain matrix to be determined such that the observer has desirable properties, such as generating an estimate $\hat{\mathbf{w}}(t)$ that can track (or converge to) $\mathbf{w}(t)$ asymptotically in the face of unknown disturbances.

Although (11) and (12) provide a more intuitive form for a Luenberger-like observer, a linear and nonlinear UIO can be expressed in an alternate form as [31, 52],

$$\begin{cases} \dot{\mathbf{q}} = \mathbf{f}_q(\mathbf{q}, \mathbf{y}), & \mathbf{q}(0) = \mathbf{q}_o, \\ \hat{\mathbf{w}} = \boldsymbol{\varphi}(\mathbf{q}, \mathbf{y}), & \hat{\mathbf{w}} = \begin{bmatrix} \hat{\mathbf{x}} \\ \hline \hat{\mathbf{m}} \end{bmatrix}, \end{cases} \tag{13}$$

where $\mathbf{q} \in \mathbb{R}^n$, $\mathbf{f}_q : \mathbb{R}^n \times \mathbb{R}^m \to \mathbb{R}^n$ is a smooth vector field, $\boldsymbol{\varphi}$ a smooth function, $\mathbf{f}_q(\mathbf{0}, \mathbf{0}) = \mathbf{0}$ and $\boldsymbol{\varphi}(\mathbf{0}, \mathbf{0}) = \mathbf{0}$.

Three classes of UIOs can be distinguished from the extended state estimate $\hat{\mathbf{w}}$, namely, (i) if $\hat{\mathbf{w}} = \begin{bmatrix} \hat{\mathbf{x}} \\ \hline \hat{\mathbf{m}} \end{bmatrix}$, then (13) is a *full-order UIO* that addresses the estimation of the entire system vector $\mathbf{x}$ and message vector $\mathbf{m}$; (ii) if $\hat{\mathbf{w}} = \begin{bmatrix} \hat{\mathbf{x}}_2 \\ \hline \hat{\mathbf{m}} \end{bmatrix}$, where $\hat{\mathbf{x}} = \begin{bmatrix} \mathbf{x}_1 \\ \hat{\mathbf{x}}_2 \end{bmatrix}$, $\mathbf{x}_1$ is known and $\hat{\mathbf{x}}_2$ is an estimate of $\mathbf{x}_2$, then (13) is a *reduced-order UIO* that deals with partial-state and message estimations; and (iii) if $\hat{\mathbf{w}} = \hat{\mathbf{m}}$, where the complete state vector $\mathbf{x}(t)$ is known for all t, then (13) is an UIO for only message estimation.

The design of all the three classes of UIOs discussed above for secure communication will be addressed in Section 4.

4. Unknown-input observers (UIOs) for chaotic secure communication

The analysis and design of UIOs for secure communication using a drive-response scheme in this section will be based on (10)–(13). Hence, (3) or (10) will serve as the drive system, while (11), (12) or (13) as the response system.

In the drive-response chaotic communication theory and applications, one of the most important issues is *synchronization*, which is closely related to the stability of the UIO. Synchronization is a property of the estimation error $\tilde{\mathbf{w}}$ given by

$$\tilde{\mathbf{w}} = \begin{bmatrix} \tilde{\mathbf{x}} \\ \hline \tilde{\mathbf{m}} \end{bmatrix}, \tag{14}$$

where $\tilde{\mathbf{x}} = \mathbf{x} - \hat{\mathbf{x}}$ and $\tilde{\mathbf{m}} = \mathbf{m} - \hat{\mathbf{m}}$.

Definition 1: Synchronization

The drive system (3) or (10) *and the UIO response system* (11), (12) or (13) *are said to be synchronized if the estimation error* $\tilde{\mathbf{w}}$ *given by* (14) *satisfies* $\lim_{t\to\infty} \|\tilde{\mathbf{w}}(t)\| = \lim_{t\to\infty} \|\mathbf{w}(t) - \hat{\mathbf{w}}(t)\| = 0$,

i.e., the UIO is capable of generating an estimate $\hat{\mathbf{w}}(t)$ *that tracks* $\mathbf{w}(t)$ *asymptotically as* $t \to \infty$. ∎

Remark 1: The condition $\lim_{t\to\infty} \|\tilde{\mathbf{w}}(t)\| = 0$ is similar to the design of linear and nonlinear observers where it is crucial to ensure the asymptotic stability of the observers. ∎

To proceed, the estimation error (14) satisfies, with (10) and (12),

$$\begin{aligned}\dot{\tilde{\mathbf{w}}} &= \mathbf{f}_w(\mathbf{w}) - \mathbf{f}_w(\hat{\mathbf{w}}) - \mathbf{L}_o(\cdot)[\mathbf{y} - \mathbf{h}(\hat{\mathbf{w}})] \\ &= \mathbf{f}_w(\mathbf{w}) - \mathbf{f}_w(\mathbf{w} - \tilde{\mathbf{w}}) - \mathbf{L}_o(\cdot)[\mathbf{h}(\mathbf{w}) - \mathbf{h}(\mathbf{w} - \tilde{\mathbf{w}})] \\ &\triangleq \mathbf{f}_w(\tilde{\mathbf{w}}, \mathbf{x}, \mathbf{m}, \mathbf{y}).\end{aligned} \tag{15}$$

It follows that $\tilde{\mathbf{w}} = \mathbf{0}$ is an equilibrium point of (15), i.e., $\mathbf{f}_w(\mathbf{w}) - \mathbf{f}_w(\mathbf{w}) - \mathbf{L}_o(\cdot)[\mathbf{h}(\mathbf{w}) - \mathbf{h}(\mathbf{w})] = \mathbf{f}_w(\mathbf{0}, \mathbf{x}, \mathbf{m}, \mathbf{y}) = \mathbf{0}$ for all $\mathbf{x}$, $\mathbf{m}$ and $\mathbf{y}$. Further, if a gain $\mathbf{L}_o(\cdot)$ can be found such that (15) is asymptotically stable, then $\lim_{t\to\infty}[\hat{\mathbf{w}}(t)] = \lim_{t\to\infty}[\mathbf{w}(t)]$, see for example [29].

The results above are stated in the following theorem.

Theorem 1: *Consider the error Eq.* (15) *with an equilibrium point at* $\tilde{\mathbf{w}} = \mathbf{0}$. *If a gain matrix* $\mathbf{L}_o(\cdot)$ *exists such that* (15) *is asymptotically stable,* then $\hat{\mathbf{w}}(t) \to \mathbf{w}(t)$ *as* $t \to \infty$. ∎

The next task is to determine the gain $\mathbf{L}_o(\cdot)$ so that the candidate observer (12) or (13) becomes an asymptotically or exponentially stable observer. The matrix can take on various forms depending on the type of systems being considered and/or the design techniques. For a general nonlinear system, $\mathbf{L}_o(\cdot)$ can be determined as a function of the estimate $\hat{\mathbf{x}}$, i.e., $\mathbf{L}_o(\cdot) = \mathbf{L}_o(\hat{\mathbf{x}})$ [27, 28]; for nonlinear systems under Jacobian linearization, $\mathbf{L}_o(\cdot)$ can be obtained as a constant matrix $\mathbf{L}_o$ [29, 30]; for extended Kalman-Bucy filtering using Jacobian linearization, the filter gain matrix can be approximated by its steady-state value $\mathbf{L}_o$. We shall focus on Jacobian linearization in Section 4.1 below with applications to full-order UIOs for state and message estimations using constant gain $\mathbf{L}_o(\cdot) = \mathbf{L}_o$. Section 4.2 addresses the design of reduced-order UIOs for message estimation, while the design of reduced-order UIOs for partial-state and message estimations is considered in Section 4.3.

4.1. Jacobian linearization: full-order UIO

Linearization of (3) or (10) about the equilibrium point $\mathbf{w}^o = \mathbf{0}$ yields

$$\bar{\mathbf{A}} = \left[\begin{array}{c|c} \mathbf{A} & \mathbf{B}_m\mathbf{M} \\ \hline \mathbf{\Psi} & \mathbf{A}_m \end{array}\right] \quad \text{and} \quad \bar{\mathbf{C}} = [\mathbf{C} \mid \mathbf{C}_m], \tag{16}$$

where

$$\mathbf{A} = \frac{\partial \mathbf{f}}{\partial \mathbf{x}}(\mathbf{x}, \mathbf{m})\Big|_{\mathbf{w}^o=0}, \qquad \mathbf{B}_m\mathbf{M} = \frac{\partial \mathbf{f}}{\partial \mathbf{m}}(\mathbf{x}, \mathbf{m})\Big|_{\mathbf{w}^o=0}, \quad (\mathbf{B}_m = \mathbf{B}_m(\mathbf{0})), \qquad \mathbf{\Psi} = \frac{\partial \mathbf{f}_m}{\partial \mathbf{x}}(\mathbf{m}, \mathbf{x})\Big|_{\mathbf{w}^o=0},$$

$$\mathbf{A}_m = \frac{\partial \mathbf{f}_m}{\partial \mathbf{m}}(\mathbf{m}, \mathbf{x})\Big|_{\mathbf{w}^o=0}, \quad \mathbf{C} = \frac{\partial \mathbf{h}}{\partial \mathbf{x}}(\mathbf{x}, \mathbf{m})\Big|_{\mathbf{w}^o=0}, \quad \mathbf{C}_m = \frac{\partial \mathbf{h}}{\partial \mathbf{m}}(\mathbf{x}, \mathbf{m})\Big|_{\mathbf{w}^o=0}.$$

The resulting linearized system is given by

$$\begin{cases} \left[\dfrac{\dot{\mathbf{x}}}{\dot{\mathbf{m}}}\right] = \bar{\mathbf{A}}\left[\dfrac{\mathbf{x}}{\mathbf{m}}\right], \\ \mathbf{y} = \bar{\mathbf{C}}\left[\dfrac{\mathbf{x}}{\mathbf{m}}\right]. \end{cases} \tag{17}$$

The following assumption is crucial to the construction of UIOs.

Assumption 1: Observability

The pair $[\overline{\mathbf{A}}, \overline{\mathbf{C}}]$ in (17) is observable, i.e.,

$\text{rank}[\mathcal{O}] = n + r,$

where $\mathcal{O}$ is the observability matrix

$$\mathcal{O} = \begin{bmatrix} \overline{\mathbf{C}}^T & \overline{\mathbf{A}}^T\overline{\mathbf{C}}^T & \left(\overline{\mathbf{A}}^2\right)^T\overline{\mathbf{C}}^T & \cdots & \left(\overline{\mathbf{A}}^{(n+r-1)}\right)^T\overline{\mathbf{C}}^T \end{bmatrix}. \qquad \blacksquare$$

An observer can be constructed for (17) if and only if $[\overline{\mathbf{A}}, \overline{\mathbf{C}}]$ is an observable pair. Hence when the Jacobian linearization method yields a pair $[\overline{\mathbf{A}}, \overline{\mathbf{C}}]$ that is not observable, then the Jacobian linearization method is not applicable to the system under consideration. However, other methods may work, such as feedback linearization [53, 54].

Using (17), a linear UIO for full-state and message estimation can be constructed as

$$\begin{cases} \begin{bmatrix} \dot{\hat{\mathbf{x}}} \\ \hline \dot{\hat{\mathbf{m}}} \end{bmatrix} = \overline{\mathbf{A}}\begin{bmatrix} \hat{\mathbf{x}} \\ \hline \hat{\mathbf{m}} \end{bmatrix} + \begin{bmatrix} \mathbf{L}_{1o} \\ \hline \mathbf{L}_{2o} \end{bmatrix}\left(\mathbf{y} - \overline{\mathbf{C}}\begin{bmatrix} \hat{\mathbf{x}} \\ \hline \hat{\mathbf{m}} \end{bmatrix}\right) \\ \quad = \left(\overline{\mathbf{A}} - \mathbf{L}_o\overline{\mathbf{C}}\right)\begin{bmatrix} \hat{\mathbf{x}} \\ \hline \hat{\mathbf{m}} \end{bmatrix} + \mathbf{L}_o\mathbf{y}, \quad \begin{bmatrix} \hat{\mathbf{x}}(0) \\ \hline \hat{\mathbf{m}}(0) \end{bmatrix} = \begin{bmatrix} \hat{\mathbf{x}}_o \\ \hline \hat{\mathbf{m}}_o \end{bmatrix}, \\ \mathbf{y} = \overline{\mathbf{C}}\begin{bmatrix} \mathbf{x} \\ \hline \mathbf{m} \end{bmatrix}, \end{cases} \tag{18}$$

where $\mathbf{L}_o = \begin{bmatrix} \mathbf{L}_{1o} \\ \hline \mathbf{L}_{2o} \end{bmatrix}$ is the constant UIO gain matrix to be determined. Note that $\mathbf{L}_{1o} \in \mathbb{R}^{n \times m}$ and $\mathbf{L}_{2o} \in \mathbb{R}^{r \times m}$, and (18) is simply a Luenberger observer [57]. Since $[\overline{\mathbf{A}}, \overline{\mathbf{C}}]$ is an observable pair by Assumption 1, then $\mathbf{L}_o$can be determined, for example, by pole-placement, such that $\left(\overline{\mathbf{A}} - \mathbf{L}_o\overline{\mathbf{C}}\right)$ is Hurwitz, i.e., all the eigenvalues of $\left(\overline{\mathbf{A}} - \mathbf{L}_o\overline{\mathbf{C}}\right)$ are located in the open left half-complex plane.

Using (17) and (18), the estimation errors $\tilde{\mathbf{x}} = \mathbf{x} - \hat{\mathbf{x}}$ and $\tilde{\mathbf{m}} = \mathbf{m} - \hat{\mathbf{m}}$ satisfy

$$\begin{bmatrix} \dot{\tilde{\mathbf{x}}} \\ \hline \dot{\tilde{\mathbf{m}}} \end{bmatrix} = \left(\overline{\mathbf{A}} - \mathbf{L}_o\overline{\mathbf{C}}\right)\begin{bmatrix} \tilde{\mathbf{x}} \\ \hline \tilde{\mathbf{m}} \end{bmatrix}, \quad \begin{bmatrix} \tilde{\mathbf{x}}(0) \\ \hline \tilde{\mathbf{m}}(0) \end{bmatrix} = \begin{bmatrix} \tilde{\mathbf{x}}_o \\ \hline \tilde{\mathbf{m}}_o \end{bmatrix}, \tag{19}$$

which is exponentially stable, i.e., $\lim_{t\to\infty}[\tilde{\mathbf{x}}(t)] = \mathbf{0}$ and $\lim_{t\to\infty}[\tilde{\mathbf{m}}(t)] = \mathbf{0}$ exponentially for all $\tilde{\mathbf{x}}(0)$ and $\tilde{\mathbf{m}}(0)$. It follows that $\hat{\mathbf{x}}(t) \to \mathbf{x}(t)$ and $\hat{\mathbf{m}}(t) \to \mathbf{m}(t)$ exponentially.

Once a constant $\mathbf{L}_o$ has been determined, it can then be substituted into (12), whereby the resulting nonlinear UIO has the form

$$\begin{cases} \dot{\hat{\mathbf{w}}} = \mathbf{f}_w(\hat{\mathbf{w}}) + \mathbf{L}_o[\mathbf{y} - \mathbf{h}(\hat{\mathbf{w}})], \quad \hat{\mathbf{w}}(0) = \hat{\mathbf{w}}_o, \\ \mathbf{y} = \mathbf{h}(\mathbf{w}), \end{cases} \tag{20}$$

where $\hat{\mathbf{w}}(0)$ is an arbitrary initial condition. Further, (15) becomes

$$\begin{aligned}\dot{\tilde{\mathbf{w}}} &= \mathbf{f}_w(\mathbf{w}) - \mathbf{f}_w(\hat{\mathbf{w}}) - \mathbf{L}_o[\mathbf{y} - \mathbf{h}(\hat{\mathbf{w}})] \\ &= \mathbf{f}_w(\mathbf{w}) - \mathbf{f}_w(\mathbf{w} - \tilde{\mathbf{w}}) - \mathbf{L}_o[\mathbf{h}(\mathbf{w}) - \mathbf{h}(\mathbf{w} - \tilde{\mathbf{w}})],\end{aligned} \tag{21}$$

which can be linearized about $\tilde{\mathbf{w}} = \mathbf{0}$ to give (19). Hence the dynamics of (21) close to the origin are well described by (19) for sufficiently small $||\hat{\mathbf{w}}(0)||$ [30].

In summary, we have the following theorem.

Theorem 2: *Let* $[\overline{\mathbf{A}}, \overline{\mathbf{C}}]$ *be an observable pair so that there exists a constant gain* $\mathbf{L}_o$ *such that* $(\overline{\mathbf{A}} - \mathbf{L}_o\overline{\mathbf{C}})$ *in* (19) *is Hurwitz. Then* (20) *is an exponentially stable dynamical system for sufficiently small* $||\hat{\mathbf{w}}(0)||$. *Further,* $\hat{x}_i(t) \to x_i(t)$ and $\hat{m}_i(t) \to m_i(t)$ imply that the drive system (10) and the UIO response system (20) are synchronized. ∎

Using (10) and (20), the overall chaotic system-based UIO for full-state and message estimations under the Jacobin linearization scheme can be implemented as

$$\begin{cases}\dot{\mathbf{w}} = \mathbf{f}_w(\mathbf{w}), \quad \mathbf{w}(0) = \mathbf{w}_o, \\ \dot{\hat{\mathbf{w}}} = \mathbf{f}_w(\hat{\mathbf{w}}) + \mathbf{L}_o[\mathbf{y} - \mathbf{h}(\hat{\mathbf{w}})], \quad \hat{\mathbf{w}}(0) = \hat{\mathbf{w}}_o, \\ \mathbf{y} = \mathbf{h}(\mathbf{x}, \mathbf{m}).\end{cases} \tag{22}$$

A block diagram for (22) is shown in **Figure 1**.

Example 1: Rossler system [47]

Consider the Rossler system with state-dependent and additive messages described by (5), with the output arranged as $\mathbf{y} = [x_1 + m_1 \quad x_3 + m_1]^T$,

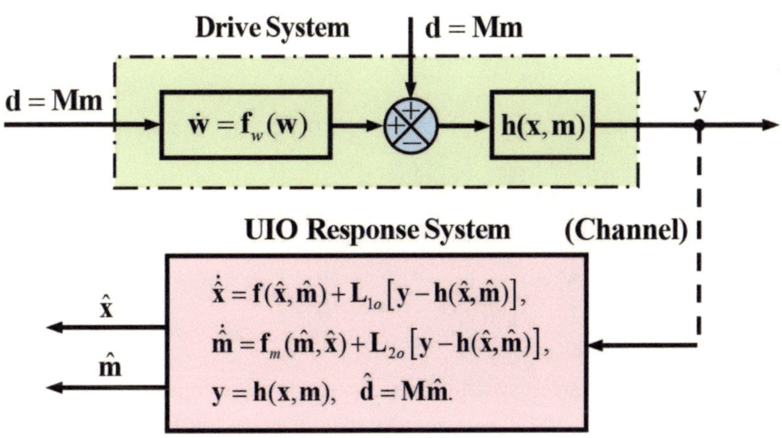

Figure 1. Chaotic secure communication system under Jacobian linearization.

$$\begin{cases} \dot{\mathbf{x}} = \begin{bmatrix} -x_2 - x_3 \\ x_1 + ax_2 \\ -(c-m_1)x_3 + x_1x_3 + (b+m_2) \end{bmatrix} = \underbrace{\begin{bmatrix} 0 & -1 & -1 \\ 1 & a & 0 \\ 0 & 0 & -c \end{bmatrix}}_{\mathbf{A}} \mathbf{x} + \underbrace{\begin{bmatrix} 0 \\ 0 \\ m_1x_3 + x_1x_3 + b \end{bmatrix}}_{\mathbf{g}(\mathbf{x},\, m_1)} + \underbrace{\begin{bmatrix} 0 & 0 \\ 0 & 0 \\ 0 & 1 \end{bmatrix}}_{\mathbf{B}_m} \mathbf{m}, \\ \dot{\mathbf{m}} = \mathbf{A}_m \mathbf{m}, \\ \mathbf{y} = \begin{bmatrix} x_1 + m_1 \\ x_3 + m_1 \end{bmatrix} = \overline{\mathbf{C}} \begin{bmatrix} \mathbf{x} \\ \mathbf{m} \end{bmatrix}, \left(\overline{\mathbf{C}} = [\mathbf{C} \quad \mathbf{C}_m], \ \mathbf{C} = \begin{bmatrix} 1 & 0 & 0 \\ 0 & 0 & 1 \end{bmatrix}, \ \mathbf{C}_m = \begin{bmatrix} 1 & 0 \\ 1 & 0 \end{bmatrix} \right). \end{cases} \tag{23}$$

The preceding equation can be expressed as

$$\begin{cases} \left[\dfrac{\dot{\mathbf{x}}}{\dot{\mathbf{m}}} \right] = \underbrace{\left[\begin{array}{c|c} \mathbf{A} & \mathbf{B}_m \\ \hline \mathbf{0} & \mathbf{A}_m \end{array} \right]}_{\overline{\mathbf{A}}} \left[\dfrac{\mathbf{x}}{\mathbf{m}} \right] + \left[\dfrac{\mathbf{g}(\mathbf{x}, m_1)}{\mathbf{0}} \right], \\ \mathbf{y} = \begin{bmatrix} x_1 + m_1 \\ x_3 + m_1 \end{bmatrix} = \overline{\mathbf{C}} \left[\dfrac{\mathbf{x}}{\mathbf{m}} \right], \end{cases} \tag{24}$$

where $\mathbf{A}_m = \mathbf{0}$ for simplicity. Note that $\left[\overline{\mathbf{A}}, \overline{\mathbf{C}}\right]$ is an observable pair for all $\mathbf{A}_m$.

It can be shown that the Rossler system $\dot{\mathbf{x}} = \mathbf{f}(\mathbf{x})$ given by (4) has two equilibrium points, for $c^2 - 4ab \geq 0$,

$$\mathbf{x}_1^o = \begin{bmatrix} \frac{1}{2}\left(c + \sqrt{c^2 - 4ab}\right) \\ \frac{1}{2a}\left(-c - \sqrt{c^2 - 4ab}\right) \\ \frac{1}{2a}\left(c + \sqrt{c^2 - 4ab}\right) \end{bmatrix} = \begin{bmatrix} 5.693 \\ -28.465 \\ 28.465 \end{bmatrix}, \quad \mathbf{x}_2^o = \begin{bmatrix} \frac{1}{2}\left(c - \sqrt{c^2 - 4ab}\right) \\ \frac{1}{2a}\left(-c + \sqrt{c^2 - 4ab}\right) \\ \frac{1}{2a}\left(c - \sqrt{c^2 - 4ab}\right) \end{bmatrix} = \begin{bmatrix} 0.0070262 \\ -0.035131 \\ 0.035131 \end{bmatrix}.$$

The stability status of $\mathbf{x}_1^o$ and $\mathbf{x}_2^o$ can be determined by checking the eigenvalues of the Jacobian matrices $\mathbf{A}_1^o = \frac{\partial \mathbf{f}}{\partial \mathbf{x}}(\mathbf{x}_1^o)$ and $\mathbf{A}_2^o = \frac{\partial \mathbf{f}}{\partial \mathbf{x}}(\mathbf{x}_2^o)$. We obtain,

$$\frac{\partial \mathbf{f}}{\partial \mathbf{x}}(\mathbf{x}) = \begin{bmatrix} 0 & -1 & -1 \\ 1 & a & 0 \\ x_3 & 0 & x_1 - c \end{bmatrix} \Rightarrow \mathbf{A}_1^o = \begin{bmatrix} 0 & -1 & -1 \\ 1 & a & 0 \\ 28.465 & 0 & -0.007 \end{bmatrix} \text{ and } \mathbf{A}_2^o = \begin{bmatrix} 0 & -1 & -1 \\ 1 & a & 0 \\ 0.035131 & 0 & -5.693 \end{bmatrix}.$$

It follows that $\mathbf{A}$, $\mathbf{A}_1^o$ and $\mathbf{A}_2^o$ are unstable matrices, since their eigenvalues have positive real parts. Since $\mathbf{A}$, $\mathbf{A}_1^o$ and $\mathbf{A}_2^o$ can all be used for the design of an UIO for full-state and message estimations, we shall choose $\mathbf{A}$ in the sequel. Therefore, using (18) and (24), it follows that the UIO for full-state and message estimations has the form

$$\begin{cases} \begin{bmatrix} \dot{\hat{\mathbf{x}}} \\ \hline \dot{\hat{\mathbf{m}}} \end{bmatrix} = \overline{\mathbf{A}} \begin{bmatrix} \hat{\mathbf{x}} \\ \hline \hat{\mathbf{m}} \end{bmatrix} + \begin{bmatrix} \mathbf{g}(\hat{\mathbf{x}}, \hat{m}_1) \\ \hline \mathbf{0} \end{bmatrix} + \begin{bmatrix} \mathbf{L}_{1o} \\ \hline \mathbf{L}_{2o} \end{bmatrix} \left(\mathbf{y} - \overline{\mathbf{C}} \begin{bmatrix} \hat{\mathbf{x}} \\ \hline \hat{\mathbf{m}} \end{bmatrix} \right), \\ \quad = \left(\overline{\mathbf{A}} - \mathbf{L}_o \overline{\mathbf{C}} \right) \begin{bmatrix} \hat{\mathbf{x}} \\ \hline \hat{\mathbf{m}} \end{bmatrix} + \begin{bmatrix} \mathbf{g}(\hat{\mathbf{x}}, \hat{m}_1) \\ \hline \mathbf{0} \end{bmatrix} + \mathbf{L}_o \mathbf{y}, \quad \begin{bmatrix} \hat{\mathbf{x}}(0) \\ \hline \hat{\mathbf{m}} \end{bmatrix} (0) = \begin{bmatrix} \hat{\mathbf{x}}_o \\ \hline \hat{\mathbf{m}}_o \end{bmatrix}, \\ \mathbf{y} = \overline{\mathbf{C}} \begin{bmatrix} \mathbf{x} \\ \hline \mathbf{m} \end{bmatrix}, \end{cases} \tag{25}$$

where the gain $\mathbf{L}_o$ is to be determined such that $\left(\overline{\mathbf{A}} - \mathbf{L}_o\overline{\mathbf{C}}\right)$ is Hurwitz. The next task is then to find $\mathbf{L}_o$, which may be obtained by using the pole-placement or Kalman-Bucy filter design method. We shall use the Kalman-Bucy filter technique. We note that in the design of a Kalman-Bucy filter [55, 56], the known covariance matrices of the system noise and measurement noise are given by $\mathbf{Q}$ and $\mathbf{R}$, respectively, where $\mathbf{Q}\geq\mathbf{0}$ and $\mathbf{R} > \mathbf{0}$. However, for the UIO design governed by (24) and (25), there are no system and measurement noises. Hence, the elements of the $\mathbf{Q}$ and $\mathbf{R}$ matrices can be treated as free design parameters to be chosen and adjusted such that the performance of the UIO is satisfactory. A general method for choosing $\mathbf{Q}$ and $\mathbf{R}$ is to set them as diagonal matrices $\mathbf{Q} = q_{ii}\mathbf{I}_n$ and $\mathbf{R} = r_{ii}\mathbf{I}_r$, where $\mathbf{I}_n$ and $\mathbf{I}_r$ are unit matrices, and adjust the values of the diagonal elements q_{ii} and r_{ii} until satisfactory responses are obtained. In general, given a set of $\{r_{ii}\}$, larger values of $\{q_{ii}\}$ will lead to larger observer gains that will place the observer poles deeper in the left half-complex plane.

The overall UIO for full-state and message estimations can be implemented as (see (22))

$$\begin{cases} \dot{\mathbf{x}} = \mathbf{A}\mathbf{x} + \mathbf{g}(\mathbf{x}, m_1) + \mathbf{B}_m\mathbf{m}, \quad \mathbf{x}(0) = \mathbf{x}_o, \\ \dot{\hat{\mathbf{x}}} = \mathbf{A}\hat{\mathbf{x}} + \mathbf{g}(\hat{\mathbf{x}}, \hat{m}_1) + \mathbf{B}_m\hat{\mathbf{m}} + \mathbf{L}_{1o}\left(\mathbf{y} - \overline{\mathbf{C}}\hat{\mathbf{y}}\right), \quad \hat{\mathbf{x}}(0) = \hat{\mathbf{x}}_o, \\ \dot{\hat{\mathbf{m}}} = \mathbf{A}_m\hat{\mathbf{m}} + \mathbf{L}_{2o}\left(\mathbf{y} - \overline{\mathbf{C}}\hat{\mathbf{y}}\right), \quad \hat{\mathbf{m}}(0) = \hat{\mathbf{m}}_o, \\ \mathbf{y} = \begin{bmatrix} x_1 + m_1 \\ x_3 + m_1 \end{bmatrix} = \overline{\mathbf{C}} \begin{bmatrix} \mathbf{x} \\ \mathbf{m} \end{bmatrix}, \end{cases} \tag{26}$$

where the messages m_1 and m_2 are injected into the Rossler system directly (see **Figure 1** also), thereby the message model $\dot{\mathbf{m}} = \mathbf{A}_m$ is omitted in (26); however, the model matrix $\mathbf{A}_m$ is needed in the message observer equation (third equation).

The key task now is the determination of a suitable UIO gain $\mathbf{L}_o$based on (24) that yields acceptable performance. The design can be accomplished by using Matlab's **LQR** command as

$$[\mathrm{L\ P\ Eo}] = \mathtt{lqr}(\mathrm{Ab}', \mathrm{Cb}', \mathrm{Q}, \mathrm{R}), \quad \mathrm{Lo} = \mathrm{L}',$$

where **Ab** and **Cb** denote $\overline{\mathbf{A}}$ and $\overline{\mathbf{C}}$, respectively; $\mathbf{L}_o = \mathbf{P}\overline{\mathbf{C}}^T\mathbf{R}^{-1}$; $\mathbf{Eo} = \lambda\left(\overline{\mathbf{A}} - \mathbf{L}_o\overline{\mathbf{C}}\right)$; and $\mathbf{P}$ is the solution of the algebraic Riccati equation (ARE)

$$\mathbf{0} = \mathbf{P}\overline{\mathbf{A}}^T + \overline{\mathbf{A}}P - \mathbf{P}\overline{\mathbf{C}}^T\mathbf{R}^{-1}\overline{\mathbf{C}}P + \mathbf{Q} = \mathbf{P}\overline{\mathbf{A}}^T + \overline{\mathbf{A}}P - \mathbf{L}_o\mathbf{R}\mathbf{L}_o^T + \mathbf{Q},$$

where $\mathbf{A}_m = \mathbf{0}$ in $\overline{\mathbf{A}}$ (see (24)). The parameter matrices **Q** and **R** that produced a suitable $\mathbf{L}_o$ were found to be given by, respectively, (note that the adjustment of **Q** was nontrivial),

$$\mathtt{Q} = \mathtt{diag}([0,\ 50,\ 1000, 5 * 10^7, 10^{12}]),\ \mathtt{R} = \mathtt{diag}([0.01,\ 0.01]).$$

We obtain

Lo =	0.99979	−0.99977	Eo = −1e + 005
	− 0.03355	−0.02882	−1884 + 1876.6i
	− 3688.8	3830.5	−1884 − 1876.6i
	51810	48122	−0.4 + 0.8i
	− 6.8055e + 006	7.33E + 06	−0.4 − 0.8i

Note that the eigenvalues $\lambda\left(\overline{\mathbf{A}} - \mathbf{L}_o\overline{\mathbf{C}}\right)$ are spread apart widely and have two complex conjugate poles.

The performance of the UIO is displayed in **Figures 2** and **3**. The initial conditions used in the simulations were: $\mathbf{x}(0) = \hat{\mathbf{x}}(0) = \begin{bmatrix} 0.2 & -0.4 & -0.2 \end{bmatrix}^T$ and $\hat{\mathbf{m}} = \begin{bmatrix} 0 & 0 \end{bmatrix}^T$. The signals to be estimated are: (a) a voice message $m_1(t)$ injected into the drive system at $t=$ 100, and (b) the electrocardiogram (ECG) $m_2(t)$ of a person. **Figure 2(a)** shows $m_1(t)$ and its estimate $\hat{m}_1(t)$, and **Figure 2(b)** shows $m_2(t)$ and $\hat{m}_2(t)$. The estimation errors were small, as can be seen from **Figure 2 (c)** and **(d)**, where the plots of m_1 vs. $\hat{m}_1$, and m_2 vs. $\hat{m}_2$ are displayed. The clean 45-degree trace in **Figure 2(c)** shows that the estimate $\hat{m}_1(t)$ of $m_1(t)$ is almost perfect, while **Figure 2(d)** shows that the estimation error $\tilde{m}_2(t) = m(t) - \hat{m}_2(t)$ was small. The synchronization of the drive-response systems is shown in **Figure 3(a)–(d)**, where $\{x_1(t), \hat{x}_1(t)\}$ and $\{x_2(t), \hat{x}_2(t)\}$ are shown; the clean 45-degree traces of $x_1(t)$ vs. $\hat{x}_1(t)$ and $x_2(t)$ vs. $\hat{x}_2(t)$ show that the synchronization was nearly perfect. Hence, we conclude that the overall synchronization of the drive-response systems and the message estimation ranges from satisfactory to excellent.

4.2. Reduced-order UIO for message estimation for completely known x(*t*)

The objective here is to estimate the unknown message signal vector $\mathbf{m}(t)$ by assuming that the entire state vector **x** is accessible by direct measurement, i.e., full-state measurement, and does not have to be estimated. Hence, without loss of generality, the output can be assumed to be given by $\mathbf{y} = \mathbf{x}$. This leads to the construction of a reduced-order UIO for message (disturbance) estimation. In general, a reduced-order observer based on full-state or partial-state measurement has an interesting structure and is an active area of research in the literature for system controls and disturbance estimation, see for example [24–26, 57–60]. The reduced-order UIO designed in this section for message estimation will be based on a *derived measurement* derived from $\mathbf{y} = \mathbf{x}$ and $\dot{\mathbf{y}} = \dot{\mathbf{x}}$; the results will be extended to partial-state and message estimations in Section 4.3.

Before launching into the design of UIOs for message estimation, let us consider a general disturbance estimation problem described by

$$\begin{cases} \dot{\mathbf{x}} = \mathbf{f}(\mathbf{x}) + \mathbf{B}_1(\mathbf{x})\mathbf{u} + \mathbf{B}_2(\mathbf{x})\mathbf{d}, \\ \mathbf{y} = \mathbf{x}, \end{cases} \tag{27}$$

where $\mathbf{x} \in \mathbb{R}^n$ is the state vector, $\mathbf{u} \in \mathbb{R}^\ell$ a known control input vector, $\mathbf{d} \in \mathbb{R}^r$ an unknown disturbance vector, and $\mathbf{y}$ the measured or known output vector; $\mathbf{f}(\mathbf{x})$, $\mathbf{B}_1(\mathbf{x})$ and $\mathbf{B}_2(\mathbf{x})$ are known function and matrices of compatible dimensions. The unknown disturbance $\mathbf{d}$ is assumed to be generated by

$$\dot{\mathbf{d}} = \mathbf{f}_d(\mathbf{d}, \mathbf{x}), \tag{28}$$

where $\mathbf{f}_d(\mathbf{0}, \mathbf{0}) = \mathbf{0}$.

The objective is to estimate the unknown disturbance $\mathbf{d}$ using the output $\mathbf{y} = \mathbf{x}$. The following lemma shows that $\mathbf{d}$ can be estimated based on a *derived measurement* instead of $\mathbf{y}$.

Lemma 1: Estimation of **d** based on derived measurement

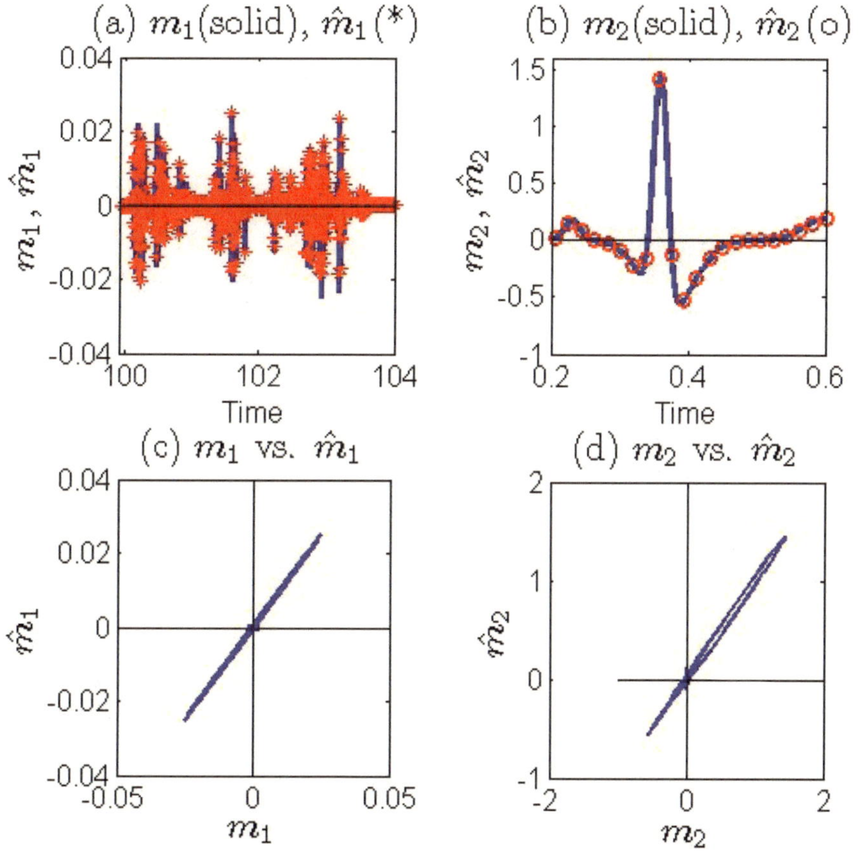

Figure 2. Responses of Rossler system: (a) m_1 and $\hat{m}_1$; (b) m_2 and $\hat{m}_2$; (c) m_1 vs. $\hat{m}_1$; and (d) m_2 vs. $\hat{m}_2$. Figures 2(c) and 2(d) indicate negligible estimation error $\tilde{m}_1 = m_1 - \hat{m}_1$ and small $\tilde{m}_2 = m_2 - \hat{m}_2$, respectively.

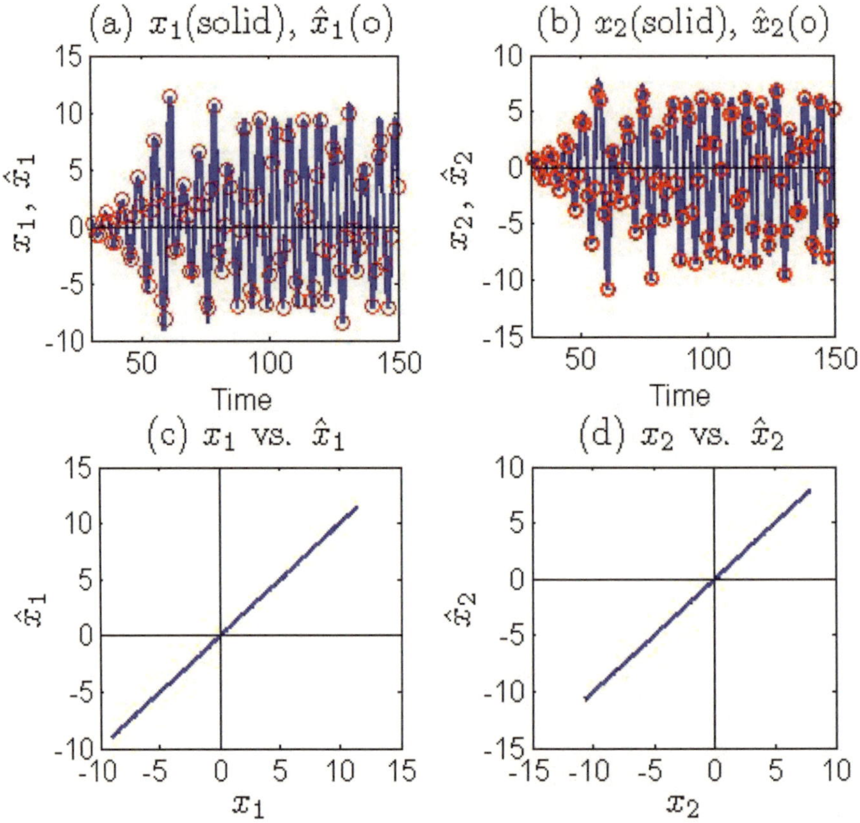

Figure 3. Responses of Rossler system: (a) x_1 and $\hat{x}_1$; (b) x_2 and $\hat{x}_2$; (c) x_1 vs. $\hat{x}_1$; and (d) x_2 vs. $\hat{x}_2$. Figures 3(c) and 3(d) indicate negligible estimation errors $\tilde{x}_1 = x_1 - \hat{x}_1$ and $\tilde{x}_2 = x_2 - \hat{x}_2$, respectively.

Consider the systems described by (27) *and* (28). *A Luenberger-like nonlinear observer can be constructed for disturbance estimation as*

$$\dot{\hat{\mathbf{d}}} = \mathbf{f}_d(\hat{\mathbf{d}}, \mathbf{x}) + \mathbf{L}_o(\mathbf{x})\left[\dot{\mathbf{x}} - \mathbf{f}(\mathbf{x}) - \mathbf{B}_1(\mathbf{x})\mathbf{u} - \mathbf{B}_2(\mathbf{x})\hat{\mathbf{d}}\right], \tag{29}$$

where $\hat{\mathbf{d}}$ *is an estimate of* $\mathbf{d}$, *and* $\mathbf{L}_o(\mathbf{x}) \in \mathbb{R}^{r \times n}$ *is the observer gain to be determined such that* $\hat{\mathbf{d}}(t) \to \mathbf{d}(t)$ *asymptotically.*

Proof: Define a *derived measurement* equation derived from the output $\mathbf{y} = \mathbf{x}$ as

$$\mathbf{y}_d = \dot{\mathbf{y}} - \mathbf{f}(\mathbf{x}) - \mathbf{B}_1(\mathbf{x})\mathbf{u}. \tag{30}$$

Since $\mathbf{x}$ is known, $\dot{\mathbf{y}} = \dot{\mathbf{x}}$ can be obtained from its time derivative; hence $\mathbf{y}_d(t)$ is known for all $t \geq 0$ for known $\mathbf{f}(\mathbf{x})$ and $\mathbf{B}_1(\mathbf{x})\mathbf{u}$. Combing (28) and (30) yields, with (27),

$$\begin{cases} \dot{\mathbf{d}} = \mathbf{f}_d(\mathbf{d}, \mathbf{x}), \\ \mathbf{y}_d = \mathbf{B}_2(\mathbf{x})\mathbf{d}, \end{cases} \tag{31}$$

which constitutes a standard form or "pattern" for constructing a nonlinear observer. Hence, a candidate Luenberger-like observer can be constructed for estimating $\mathbf{d}$ based on $\mathbf{y}_d$ (or simply by "pattern recognition") as

$$\dot{\hat{\mathbf{d}}} = \mathbf{f}_d(\hat{\mathbf{d}}, \mathbf{x}) + \mathbf{L}_o(\mathbf{x})\left[\mathbf{y}_d - \mathbf{B}_2(\mathbf{x})\hat{\mathbf{d}}\right]. \tag{32}$$

Substituting (30) into (32) yields (29). ∎

Remark 2: When $\mathbf{f}_d(\hat{\mathbf{d}}, \mathbf{x}) = \mathbf{0}$, (29) is identical to the observer proposed in Ref. [61] (Eq. (3.2), p. 44) under a versatile *disturbance observer-based control (DOBC)* technique applicable to both linear and nonlinear systems. However, it is not clear how and why the derivative term $\dot{\mathbf{x}}$ shows up in their Eq. (3.2). In contrast, the formulation in Lemma 1 based on the method of *derived measurement* provides a clear insight, specifically, it clearly shows how $\dot{\mathbf{x}}$ finds its way into (29). Furthermore, since it is, in general, difficult to access the entire system state vector $\mathbf{x}$ for measurement, the derived measurement formulation will pave the way for the design of reduced-order observers for joint partial-state and disturbance estimations (see Section 4.3) using only those state variables that are available by direct measurement, thereby extending the DOBC technique and applications. ∎

Remark 3: The presence or origin of $\dot{\mathbf{y}}$ in a linear Luenberger observer is well known in the literature [57, 58]. It occurs in the construction of reduced-order linear observers, where the elimination of $\dot{\mathbf{y}}$ leads to the design of improved or enhanced reduced-order observers. As shown in (29), the derivative $\dot{\mathbf{y}}$ also occurs in constructing enhanced reduced-order nonlinear observers. ∎

To continue further, the derivative $\dot{\mathbf{y}}$ in (29) can be eliminated by moving the term $\mathbf{L}(\mathbf{x})\dot{\mathbf{y}}$ to the left side of the equation to yield

$$\dot{\mathbf{z}} = \left[\mathbf{f}_d(\hat{\mathbf{d}}, \mathbf{x}) - \mathbf{L}_o(\mathbf{x})\mathbf{B}_2(\mathbf{x})\hat{\mathbf{d}}\right] - \mathbf{L}_o(\mathbf{x})[\mathbf{f}(\mathbf{x}) + \mathbf{B}_1(\mathbf{x})\mathbf{u}], \tag{33}$$

where $\dot{\mathbf{z}} = \dot{\hat{\mathbf{d}}} - \mathbf{L}_o(\mathbf{x})\dot{\mathbf{y}}$. Defining [61],

$$\mathbf{z} = \hat{\mathbf{d}} - \mathbf{p}(\mathbf{x})\mathbf{y} \Rightarrow \dot{\mathbf{z}} = \dot{\hat{\mathbf{d}}} - \frac{\partial \mathbf{p}(\mathbf{x})}{\partial \mathbf{x}}\dot{\mathbf{y}} \Rightarrow \mathbf{L}_o(\mathbf{x}) = \frac{\partial \mathbf{p}(\mathbf{x})}{\partial \mathbf{x}}, \tag{34}$$

where $\mathbf{p}(\mathbf{x}) \in \mathbb{R}^{r\times n}$ is to be determined. If $\mathbf{f}_d(\hat{\mathbf{d}}, \mathbf{x}) = \mathbf{0}$, then (33) can be expressed as

$$\begin{cases} \dot{\mathbf{z}} = -\mathbf{L}_o(\mathbf{x})\mathbf{B}_2(\mathbf{x})\mathbf{z} - \mathbf{L}_o(\mathbf{x})[\mathbf{B}_2(\mathbf{x})\mathbf{p}(\mathbf{x}) + \mathbf{f}(\mathbf{x}) + \mathbf{B}_1(\mathbf{x})\mathbf{u}], \\ \hat{\mathbf{d}} = \mathbf{z} + \mathbf{p}(\mathbf{x})\mathbf{y}, \end{cases} \tag{35}$$

which is identical to the enhanced observer presented in Ref. [61] (see for example, Eq. (3.5), p. 44).

We now return to message estimation in chaotic systems. We can start with (3), which can be expressed as, with full-state measurement given by $\mathbf{y} = \mathbf{x}$,

$$\begin{cases} \dot{\mathbf{x}} = \mathbf{f}(\mathbf{x}) + \mathbf{B}_m(\mathbf{x})\mathbf{M}\mathbf{m}, \\ \dot{\mathbf{m}} = \mathbf{A}_m\mathbf{m} + \mathbf{\Psi}\mathbf{x}, \\ \mathbf{y} = \mathbf{x}. \end{cases} \tag{36}$$

Since the entire state vector $\mathbf{x}$ is known for all $t \geq 0$, (36) can be rearranged as

$$\begin{cases} \dot{\mathbf{m}} = \mathbf{A}_m\mathbf{m} + \mathbf{\Psi}\mathbf{x}, \\ \mathbf{y}_d = \mathbf{B}_m(\mathbf{x})\mathbf{M}\mathbf{m}, \end{cases} \tag{37}$$

where

$$\mathbf{y}_d \triangleq \dot{\mathbf{y}} - \mathbf{f}(\mathbf{x}), \quad (\dot{\mathbf{y}} = \dot{\mathbf{x}}), \tag{38}$$

is the *derived-measurement* in the form of (30). Most importantly, $\mathbf{y}_d$ can serve as the output equation for $\dot{\mathbf{m}} = \mathbf{A}_m\mathbf{m} + \mathbf{\Psi}\mathbf{x}$, so that (37) provides a standard form or pattern for observer design. Accordingly, a candidate Luenberger-like observer can be constructed based on $\mathbf{y}_d$ as

$$\begin{aligned} \dot{\hat{\mathbf{m}}} &= \mathbf{A}_m\hat{\mathbf{m}} + \mathbf{\Psi}\mathbf{x} + \mathbf{L}_o(\mathbf{x})\left[\mathbf{y}_d - \mathbf{B}_m(\mathbf{x})\mathbf{M}\hat{\mathbf{m}}\right] \\ &= [\mathbf{A}_m - \mathbf{L}_o(\mathbf{x})\mathbf{B}_m(\mathbf{x})\mathbf{M}]\hat{\mathbf{m}} + \mathbf{\Psi}\mathbf{x} + \mathbf{L}_o(\mathbf{x})[\dot{\mathbf{y}} - \mathbf{f}(\mathbf{x})], \quad \hat{\mathbf{m}}(0) = \hat{\mathbf{m}}_o, \end{aligned} \tag{39}$$

where $\mathbf{L}_o(\mathbf{x}) \in \mathbb{R}^{r \times n}$ is the observer gain matrix to be determined.

To proceed, it follows from (37) and (39) that the estimation error defined by $\tilde{\mathbf{m}} = \mathbf{m} - \hat{\mathbf{m}}$ satisfies

$$\dot{\tilde{\mathbf{m}}} = [\mathbf{A}_m - \mathbf{L}_o(\mathbf{x})\mathbf{B}_m(\mathbf{x})\mathbf{M}]\tilde{\mathbf{m}}, \quad \tilde{\mathbf{m}}(0) = \tilde{\mathbf{m}}_o, \tag{40}$$

which shows that if $\mathbf{L}_o(\mathbf{x})$ is a suitable stabilizing gain, then $\tilde{\mathbf{m}}(t)$ can be made to converge to zero asymptotically for arbitrary $\tilde{\mathbf{m}}(0)$, thereby $\hat{\mathbf{m}}(t) \to \mathbf{m}(t)$.

The results above are summarized in the following theorem.

Theorem 3: *Consider (36)–(39). If there exists a gain $\mathbf{L}_o(\mathbf{x})$ such that (40) is asymptotically stable for all $\mathbf{x}$, then $\hat{\mathbf{m}}(t) \to \mathbf{m}(t)$ asymptotically.* ∎

Note that since $\mathbf{B}_m(\mathbf{x})$ is a function of $\mathbf{x}$, it complicates the determination of $\mathbf{L}_o(\mathbf{x})$ to achieve asymptotic stability. However, if $\mathbf{B}_m(\mathbf{x}) = \mathbf{B}_m$, where $\mathbf{B}_m$ is a constant matrix, then $\mathbf{L}_o(\mathbf{x})$ can be determined as a constant $\mathbf{L}_o$, and can be computed by simple methods, such as pole placement, provided that $[\mathbf{A}_m, \mathbf{B}_m]$ is an observable pair (see Example 2 below).

Using (36), it follows that (35) takes on the form,

$$\begin{cases} \dot{\mathbf{z}} = [\mathbf{A}_m - \mathbf{L}_o(\mathbf{x})\mathbf{B}_m(\mathbf{x})\mathbf{M}]\mathbf{z} + \mathbf{\Psi}\mathbf{x} + \mathbf{A}_m\mathbf{p}(\mathbf{x}) - \mathbf{L}_o(\mathbf{x})[\mathbf{B}_m(\mathbf{x})\mathbf{M}\mathbf{p}(\mathbf{x}) + \mathbf{f}(\mathbf{x})], \\ \hat{\mathbf{m}} = \mathbf{z} + \mathbf{p}(\mathbf{x}). \end{cases} \tag{41}$$

A main task in applying (41) is the determination of $\mathbf{p}(\mathbf{x})$. If we set $\mathbf{p}(\mathbf{x})$as a linear function of $\mathbf{x}$, i.e.,

$$\mathbf{p}(\mathbf{x}) = \mathbf{L}_o\mathbf{x}, \quad (\mathbf{x} = \mathbf{y}), \tag{42}$$

where $\mathbf{L}_o$ is a constant matrix, then we obtain from (34), $\mathbf{L}(\mathbf{x}) = \mathbf{L}_o$. Further, if $\mathbf{B}_m(\mathbf{x}) = \mathbf{B}_m$ and $[\mathbf{A}_m, \mathbf{B}_m\mathbf{M}]$ is an observable pair, then $\mathbf{L}_o$ can be determined readily by, for example, the pole-placement method, such that $(\mathbf{A}_m - \mathbf{L}_o\mathbf{B}_m\mathbf{M})$ is Hurwitz. Moreover, (40) becomes,

$$\dot{\tilde{\mathbf{m}}} = (\mathbf{A}_m - \mathbf{L}_o\mathbf{B}_m\mathbf{M})\tilde{\mathbf{m}}, \quad \tilde{\mathbf{m}}(0) = \tilde{\mathbf{m}}_o, \tag{43}$$

which shows that $\tilde{\mathbf{m}}(t) \to \mathbf{0}$, thereby $\hat{\mathbf{m}}(t) \to \mathbf{m}(t)$ exponentially for arbitrary $\tilde{\mathbf{m}}(0)$. In addition, in this case, the enhanced UIO (41) reduces to

$$\begin{cases} \dot{\mathbf{z}} = (\mathbf{A}_m - \mathbf{L}_o\mathbf{B}_m\mathbf{M})\mathbf{z} + [(\mathbf{A}_m - \mathbf{L}_o\mathbf{B}_m\mathbf{M})\mathbf{L}_o\mathbf{x} + \mathbf{\Psi}\mathbf{x} - \mathbf{L}_o\mathbf{f}(\mathbf{x})], \\ \hat{\mathbf{m}} = \mathbf{z} + \mathbf{L}_o\mathbf{y}. \end{cases} \tag{44}$$

It can be shown that the preceding equation can be obtained by using the linearized system (17) and setting $\mathbf{p}(\mathbf{x}) = \mathbf{L}_o\mathbf{x} \Rightarrow \mathbf{L}_o = \partial\mathbf{p}(\mathbf{x})/\partial\mathbf{x}$.

Once a suitable gain has been determined, such as $\mathbf{L}_o(\mathbf{x}) = \mathbf{L}_o$, it can then be substituted into (41), and the overall chaotic system-based UIO for message estimation can be implemented as, with (36),

$$\begin{cases} \dot{\mathbf{x}} = \mathbf{f}(\mathbf{x}) + \mathbf{B}_m(\mathbf{x})\mathbf{M}\mathbf{m}, \quad \mathbf{x}(0) = \mathbf{x}_o, \\ \dot{\mathbf{m}} = \mathbf{A}_m\mathbf{m} + \mathbf{\Psi}\mathbf{x}, \quad \mathbf{m}(0) = \mathbf{m}_o, \\ \dot{\mathbf{z}} = (\mathbf{A}_m - \mathbf{L}_o\mathbf{B}_m(\mathbf{x})\mathbf{M})\mathbf{z} + [(\mathbf{A}_m - \mathbf{L}_o\mathbf{B}_m(\mathbf{x})\mathbf{M})\mathbf{L}_o\mathbf{x} + \mathbf{\Psi}\mathbf{x} - \mathbf{L}_o\mathbf{f}(\mathbf{x})], \quad \mathbf{z}(0) = \mathbf{z}_o, \\ \hat{\mathbf{m}} = \mathbf{z} + \mathbf{L}_o\mathbf{y}, \\ \mathbf{y} = \mathbf{x}. \end{cases} \tag{45}$$

We remark that the UIO governed by the third equation in (45) is a nonlinear observer with its gain $\mathbf{L}_o(\mathbf{x})$ replaced by a constant $\mathbf{L}_o$. Other methods may be used to determine a suitable $\mathbf{L}_o$, such as linear matrix inequality (LMI), see for example Ref. [34].

Example 2: Genesio-Tesi system [49]

Consider the Genesio-Tesi system described by (7) with additive messages and output $\mathbf{y} = \mathbf{x}$

$$\begin{cases} \dot{\mathbf{x}} = \begin{bmatrix} x_2 \\ x_3 + m_1 \\ -cx_1 - bx_2 - ax_3 + x_1^2 + m_2 \end{bmatrix} = \mathbf{f}(\mathbf{x}) + \underbrace{\begin{bmatrix} 0 & 0 \\ 1 & 0 \\ 0 & 1 \end{bmatrix}}_{\mathbf{B}_m} \begin{bmatrix} m_1 \\ m_2 \end{bmatrix}, \\ \dot{\mathbf{m}} = \mathbf{A}_{\mathrm{m}}\mathbf{m}, \\ \mathbf{y} = \mathbf{x}. \end{cases} \tag{46}$$

Using (37) with $\mathbf{\Psi} = \mathbf{0}$, the preceding equation can be arranged in the form of an LTI system as

$$\begin{cases} \dot{\mathbf{m}} = \mathbf{A}_m\mathbf{m}, \\ \mathbf{y}_d = \mathbf{B}_m\mathbf{m}, \end{cases} \tag{47}$$

where $\mathbf{y}_d = \dot{\mathbf{y}} - \mathbf{f}(\mathbf{x})$ is the derived measurement, and it can be shown that $[\mathbf{A}_m, \mathbf{B}_m]$ is an observable pair for all $\mathbf{A}_m$, i.e., $\text{rank}\begin{bmatrix} \mathbf{B}_m^T & \mathbf{A}_m^T \mathbf{B}_m^T \end{bmatrix} = 2$.

An observer for (47) can be constructed as

$$\begin{aligned} \dot{\hat{\mathbf{m}}} &= \mathbf{A}_m \hat{\mathbf{m}} + \mathbf{L}_o(\mathbf{y}_d - \mathbf{L}_o \mathbf{B}_m \hat{\mathbf{m}}) \\ &= (\mathbf{A}_m - \mathbf{L}_o \mathbf{B}_m)\hat{\mathbf{m}} + \mathbf{L}_o[\dot{\mathbf{y}} - \mathbf{f}(\mathbf{x})], \end{aligned} \tag{48}$$

which is obtainable from (39). Since $[\mathbf{A}_m, \mathbf{B}_m]$ is an observable pair, a constant gain $\mathbf{L}_o$ can be determined such that $(\mathbf{A}_m - \mathbf{L}_o \mathbf{B}_m)$ is Hurwitz. Further, eliminating the derivative term $\mathbf{L}_o \dot{\mathbf{y}}$ in (48) yields

$$\begin{cases} \dot{\mathbf{z}} = (\mathbf{A}_m - \mathbf{L}_o \mathbf{B}_m)\mathbf{z} + [(\mathbf{A}_m - \mathbf{L}_o \mathbf{B}_m)\mathbf{L}_o \mathbf{y} - \mathbf{L}_o \mathbf{f}(\mathbf{x})], \\ \hat{\mathbf{m}} = \mathbf{z} + \mathbf{L}_o \mathbf{y}. \end{cases} \tag{49}$$

To determine the gain $\mathbf{L}_o$, let the poles of $(\mathbf{A}_m - \mathbf{L}_o \mathbf{B}_m)$ be selected as $\mathbf{po} = -\begin{bmatrix} 61 & 32 \end{bmatrix}$. Using Matlab's pole-placement command,

```
Lo=place(Am',Bm',po)',
```

we obtain, for $\mathbf{A}_m = \mathbf{0}$,

Lo = 0	61	0
0	0	32

The final result for implementation can be obtained by combing Eqs. (46) and (49) as

$$\begin{cases} \dot{\mathbf{x}} = \mathbf{f}(\mathbf{x}) + \mathbf{B}_m \mathbf{m}, \quad \mathbf{x}(0) = \mathbf{x}_o, \\ \dot{\mathbf{z}} = (\mathbf{A}_m - \mathbf{L}_o \mathbf{B}_m)\mathbf{z} + [(\mathbf{A}_m - \mathbf{L}_o \mathbf{B}_m)\mathbf{L}_o \mathbf{y} - \mathbf{L}_o \mathbf{f}(\mathbf{x})], \quad \mathbf{z}(0) = \mathbf{z}_o, \\ \hat{\mathbf{m}} = \mathbf{z} + \mathbf{L}_o \mathbf{y}, \\ \mathbf{y} = \mathbf{x}. \end{cases} \tag{50}$$

Since $m_1(t)$ and $m_2(t)$ are injected directly into the Genesio-Tesi system (46), the message model $\dot{\mathbf{m}} = \mathbf{A}_m \mathbf{m}$ is not needed and is omitted in (50); however, the model matrix $\mathbf{A}_m$ is required in the estimation equation (second equation in (50)). The signal $m_1(t)$ is the nine-term Fourier series of a square wave, and $m_2(t)$ is a mix signal consisting of a trapezoid, sine wave, and ramp and exponential functions. It would be difficult to generate these rather complicated signals, in particular $m_2(t)$, by using the simple model $\dot{\mathbf{m}} = \mathbf{A}_m \mathbf{m}$, and/or a more general model $\dot{\mathbf{m}} = \mathbf{A}_m \mathbf{m} + \delta$ proposed in Refs. [24–26], where the elements $\delta_i(t)$ of $\delta(t)$ are unknown sequences of random delta functions. For simulation studies, the mix signal $m_2(t)$ can easily be generated by the following Matlab codes and injected into (50):

Mix signal $m_2(t)$:

```
m2=0.05*t*((t>0)&(t<10))+0.5*((t>=10)&(t<=20))
    -0.05*(t-30)*((t>20)&(t<=30))+0.25*sin(t-30)*((t>=30)&(t<58.27))
```

```
+(1/20)*(t-58.27)*((t>=58.27)&(t<78.27))
+1*exp(-0.2*(t-78.27))*((t>=78.27)&(t<200)).
```

The performance of the UIO is displayed in **Figures 4** and **5**. The initial condition of the Genesio-Tesi system used in the simulation was $\mathbf{x}(0) = [0.2 \quad -0.4 \quad -0.2]^T$, while $\mathbf{z}(0)$ was calculated by using $\mathbf{z}(0) = \hat{\mathbf{m}}(0) - \mathbf{L}_o\mathbf{y}(0) = -\mathbf{L}_o\mathbf{x}(0)$, which yields $\mathbf{z}(0) = [24.4 \quad 6.4]^T$ where $\hat{\mathbf{m}}(0) = \mathbf{0}$. **Figure 4(a)** shows $m_1(t)$ and its estimate $\hat{m}_1(t)$, and **Figure 4(b)** exhibits $m_2(t)$ and $\hat{m}_2(t)$. The estimation errors were negligible, as can be seen from **Figure 4(c)** and **(d)**, where the plots of $m_1(t)$ vs. $\hat{m}_1(t)$, and $m_2(t)$ vs. $\hat{m}_2(t)$ are displayed. Note also the Gibb's phenomenon (the "twin-towers") in **Figure 4(a)**. The Genesio-Tesi attractor is shown in **Figure 5**. We conclude that the performance of the reduced-order UIO for message estimation was satisfactory.

4.3. Reduced-order UIO for partial-state and message estimations

The objective in this section is to extend the design of reduced-order UIO for message estimation to the design of UIO for joint partial-state and message estimations. The results obtained are believed to be new and novel.

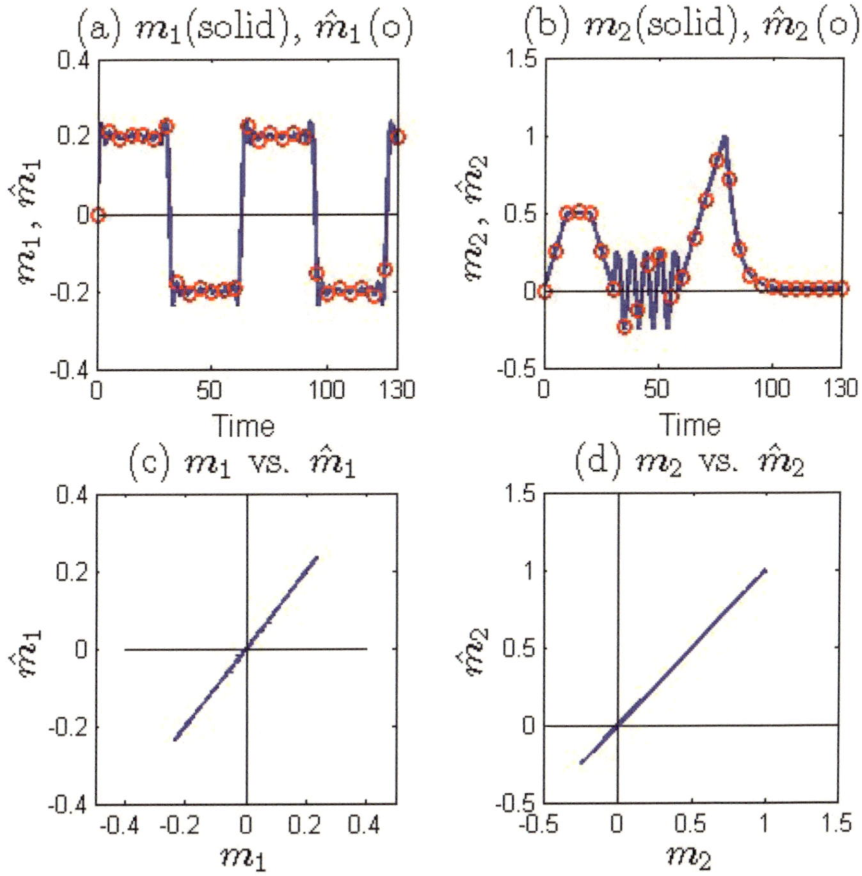

Figure 4. Responses of Genesio-Tesi system: (a) m_1 and $\hat{m}_1$; (b) m_2 and $\hat{m}_2$; (c) m_1 vs. $\hat{m}_1$; and (d) m_2 vs. $\hat{m}_2$. Figures 4(c) and 4(d) indicate negligible estimation error $\tilde{m}_1 = m_1 - \hat{m}_1$ and $\tilde{m}_2 = m_2 - \hat{m}_2$, respectively.

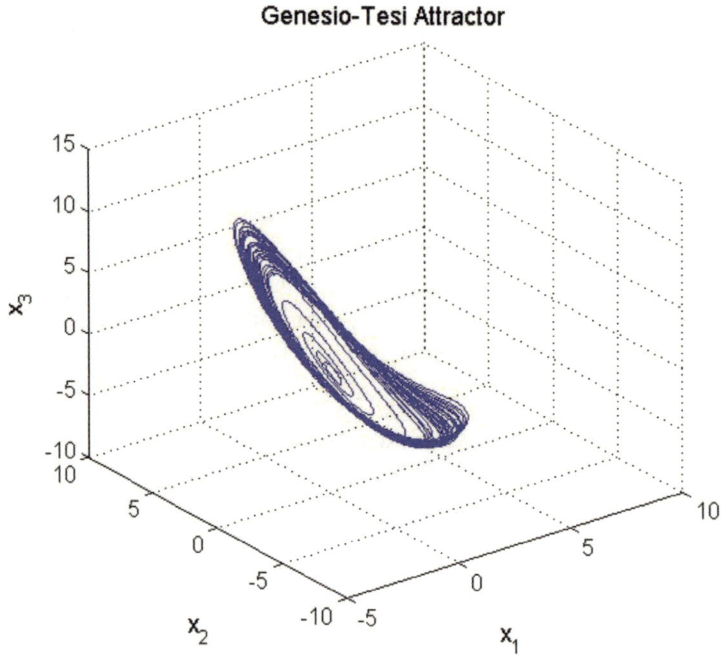

Figure 5. Genesio-Tesi attractor obtained by using (46).

Consider a general nonlinear system described by (3) which is expressed here without the output as (see also (1) and (2))

$$\begin{cases} \dot{\mathbf{x}} = \mathbf{f}(\mathbf{x}) + \mathbf{B}_m(\mathbf{x})\mathbf{M}\mathbf{m}, \\ \dot{\mathbf{m}} = \mathbf{A}_m\mathbf{m}, \end{cases} \tag{51}$$

where $\mathbf{\Psi} = \mathbf{0}$. The design will be based on a derived measurement formulation.

Let

$$\mathbf{x} = \begin{bmatrix} \mathbf{x}_1 \\ \mathbf{x}_2 \end{bmatrix}, \qquad \mathbf{w} \triangleq \begin{bmatrix} \mathbf{x}_2 \\ \mathbf{m} \end{bmatrix}, \qquad \mathbf{y} = \mathbf{x}_1, \tag{52}$$

where $\mathbf{x}_1 \in \mathbb{R}^m$ and $\mathbf{x}_2 \in \mathbb{R}^{n-m}$ are, respectively, accessible and inaccessible for direct measurement, and $\mathbf{y}$ is the output. Using (52), we assume that (51) can be partitioned as

$$\begin{cases} \dot{\mathbf{x}}_1 = \mathbf{f}_1(\mathbf{y}) + \mathbf{B}_{1m}(\mathbf{y})\mathbf{M}\mathbf{m}, \quad (\mathbf{y} = \mathbf{x}_1), \\ \dot{\mathbf{x}}_2 = \mathbf{f}_2(\mathbf{y}, \mathbf{x}_2, \mathbf{m}) + \mathbf{B}_{2m}(\mathbf{y}, \mathbf{x}_2)\mathbf{M}\mathbf{m}, \\ \dot{\mathbf{m}} = \mathbf{A}_m\mathbf{m}, \end{cases} \tag{53}$$

which can be rearranged to give

$$\begin{cases} \begin{bmatrix} \dot{\mathbf{x}}_2 \\ \dot{\mathbf{m}} \end{bmatrix} = \begin{bmatrix} \mathbf{f}_2(\mathbf{y}, \mathbf{x}_2, \mathbf{m}) \\ \mathbf{A}_m \mathbf{m} \end{bmatrix} + \begin{bmatrix} \mathbf{B}_{2m}(\mathbf{y}, \mathbf{x}_2)\mathbf{M}\mathbf{m} \\ \mathbf{0} \end{bmatrix}, \\ \mathbf{y}_d = \mathbf{B}_{1m}(\mathbf{y})\mathbf{M}\mathbf{m}, \end{cases} \tag{54}$$

where $\mathbf{y}_d = \dot{\mathbf{y}} - \mathbf{f}_1(\mathbf{y})$ denotes the derived measurement.

Eq. (54) constitutes a standard form that can be used to construct an observer for estimating the inaccessible partial-state $\mathbf{x}_2$ and the unknown message $\mathbf{m}$ based on the derived measurement $\mathbf{y}_d$. Hence, a candidate Luenberger-like observer can be constructed as

$$\dot{\hat{\mathbf{w}}} = \begin{bmatrix} \mathbf{f}_2(\mathbf{y}, \hat{\mathbf{x}}_2, \hat{\mathbf{m}}) \\ \mathbf{A}_m \hat{\mathbf{m}} \end{bmatrix} + \begin{bmatrix} \mathbf{B}_{2m}(\mathbf{y}, \hat{\mathbf{x}}_2)\mathbf{M}\hat{\mathbf{m}} \\ \mathbf{0} \end{bmatrix} + \begin{bmatrix} \mathbf{L}_{1o}(\cdot) \\ \mathbf{L}_{2o}(\cdot) \end{bmatrix} \left[\mathbf{y}_d - \mathbf{B}_{1m}(\mathbf{y})\mathbf{M}\hat{\mathbf{m}}\right], \tag{55}$$

where $\mathbf{L}_{1o}(\cdot)$ and $\mathbf{L}_{2o}(\cdot)$ are the gain matrices to be determined such that the observer has desirable performance characteristics, in particular, $\hat{\mathbf{x}}_2(t) \to \mathbf{x}_2(t)$ and $\hat{\mathbf{m}}(t) \to \mathbf{m}(t)$ as $t \to \infty$.

Remark 4: In (39), the reduced-order UIO was derived using the output $\mathbf{y} = \mathbf{x}$, while the reduced-order UIO (55) above was constructed by using $\mathbf{y} = \mathbf{x}_1$ with $\mathbf{x}_1$ serving the role of $\mathbf{x}$. Hence (55) is an extension of the DOBC technique, and is now applicable to partial-state and message estimations by using only $\mathbf{x}_1$ instead of the entire state $\mathbf{x}$.

The estimation error $\tilde{\mathbf{w}} = \left[\tilde{\mathbf{x}}_2^T \quad \tilde{\mathbf{m}}^T\right]^T$, where $\tilde{\mathbf{x}}_2 = \mathbf{x}_2 - \hat{\mathbf{x}}_2$ and $\tilde{\mathbf{m}}_2 = \mathbf{m} - \hat{\mathbf{m}}$, satisfies, with (54) and (55),

$$\dot{\tilde{\mathbf{w}}} = \begin{bmatrix} \mathbf{f}_2(\mathbf{y}, \mathbf{x}_2, \mathbf{m}) - \mathbf{f}_2(\mathbf{y}, \hat{\mathbf{x}}_2, \hat{\mathbf{m}}) \\ \mathbf{A}_m \mathbf{m} - \mathbf{A}_m \hat{\mathbf{m}} \end{bmatrix} + \begin{bmatrix} \mathbf{B}_{2m}(\mathbf{y}, \mathbf{x}_2)\mathbf{M}\hat{\mathbf{m}} - \mathbf{B}_{2m}(\mathbf{y}, \hat{\mathbf{x}}_2)\mathbf{M}\hat{\mathbf{m}} \\ \mathbf{0} \end{bmatrix} - \begin{bmatrix} \mathbf{L}_{1o}(\cdot) \\ \mathbf{L}_{2o}(\cdot) \end{bmatrix} \left[\mathbf{y}_d - \mathbf{B}_{1m}(\mathbf{y})\mathbf{M}\hat{\mathbf{m}}\right]. \tag{56}$$

The preceding error equation is a version of (15). Hence from Theorem 1, the origin $(\mathbf{0}, \mathbf{0})$ is an equilibrium point of the unforced equation in (56) for all $\mathbf{y} = \mathbf{x}_1$. Further, $\hat{\mathbf{x}}_2(t) \to \mathbf{x}_2(t)$ and $\hat{\mathbf{m}}(t) \to \mathbf{m}(t)$ if $\mathbf{L}_{1o}(\cdot)$ and $\mathbf{L}_{2o}(\cdot)$ are stabilizing gains.

The next task is to eliminate $\dot{\mathbf{y}}$ in $\mathbf{y}_d = \dot{\mathbf{y}} - \mathbf{f}_1(\mathbf{y})$ in (55) by moving $\mathbf{L}_o(\cdot)\dot{\mathbf{y}}$ to the left side of the equation and defining

$$\dot{\mathbf{z}} = \dot{\hat{\mathbf{w}}} - \mathbf{L}_o(\cdot)\dot{\mathbf{y}}. \tag{57}$$

Choosing

$$\mathbf{z} = \hat{\mathbf{w}} - \mathbf{p}(\mathbf{x}_1) \Rightarrow \dot{\mathbf{z}} = \dot{\hat{\mathbf{w}}} - \frac{\partial \mathbf{p}(\mathbf{x}_1)}{\partial \mathbf{x}_1}\dot{\mathbf{x}}_1, \quad (\dot{\mathbf{x}}_1 = \dot{\mathbf{y}}), \tag{58}$$

where $\mathbf{p}(\mathbf{x}_1) = \begin{bmatrix} \mathbf{p}_1(\mathbf{x}_1) \\ \mathbf{p}_2(\mathbf{x}_1) \end{bmatrix}$, $\mathbf{p}_1(\mathbf{x}_1) \in \mathbb{R}^{(n-m)\times(n-m)}$ and $\mathbf{p}_2(\mathbf{x}_1) \in \mathbb{R}^{r\times(n-m)}$ are to be determined. It follows that

$$\mathbf{L}_{1o}(\mathbf{x}_1) = \frac{\partial \mathbf{p}_1(\mathbf{x}_1)}{\partial \mathbf{x}_1} \quad \text{and} \quad \mathbf{L}_{2o}(\mathbf{x}_1) = \frac{\partial \mathbf{p}_2(\mathbf{x}_1)}{\partial \mathbf{x}_1} \tag{59}$$

Using (55), (57), (58) and (59), can be expressed as

$$\begin{cases} \dot{\mathbf{z}} = \begin{bmatrix} \mathbf{f}_2(\mathbf{y}, \hat{\mathbf{x}}_2, \hat{\mathbf{m}}) \\ \mathbf{A}_m \hat{\mathbf{m}} \end{bmatrix} + \begin{bmatrix} \mathbf{B}_{2m}(\mathbf{y}, \hat{\mathbf{x}}_2)\mathbf{M}\hat{\mathbf{m}} \\ \mathbf{0} \end{bmatrix} - \begin{bmatrix} \mathbf{L}_{1o}(\mathbf{x}_1) \\ \mathbf{L}_{2o}(\mathbf{x}_1) \end{bmatrix} [\mathbf{f}_1(\mathbf{y}) + \mathbf{B}_{1m}(\mathbf{y})\mathbf{M}\hat{\mathbf{m}}], \\ \hat{\mathbf{w}} = \mathbf{z} + \mathbf{p}(\mathbf{x}_1), \end{cases} \tag{60}$$

which can further be reduced to a form given by, for example (44), once the specific structure of the chaotic system under consideration is known and $\mathbf{p}(\mathbf{x}_1)$ has been determined (see Example 3 for more details).

Using (51) and (60), the main results for the construction of UIO for partial-state and message estimations are stated in the following theorem.

Theorem 4: *Consider the augmented system (54), where* $\mathbf{y}_d = \mathbf{B}_{1m}(\mathbf{y})\mathbf{M}\mathbf{m}$ *is the derived measurement. A candidate UIO for partial-state and message estimations is given by*

$$\begin{cases} \dot{\mathbf{z}} = \begin{bmatrix} \mathbf{f}_2(\mathbf{y}, \hat{\mathbf{x}}_2, \hat{\mathbf{m}}) \\ \mathbf{A}_m \hat{\mathbf{m}} \end{bmatrix} + \begin{bmatrix} \mathbf{B}_{2m}(\mathbf{y}, \hat{\mathbf{x}}_2)\mathbf{M}\hat{\mathbf{m}} \\ \mathbf{0} \end{bmatrix} - \begin{bmatrix} \mathbf{L}_{1o}(\mathbf{x}_1) \\ \mathbf{L}_{2o}(\mathbf{x}_1) \end{bmatrix} [\mathbf{f}_1(\mathbf{y}) + \mathbf{B}_{1m}(\mathbf{y})\mathbf{M}\hat{\mathbf{m}}], \\ \hat{\mathbf{w}} = \mathbf{z} + \mathbf{p}(\mathbf{x}_1), \\ \mathbf{y} = \mathbf{x}_1. \end{cases} \tag{61}$$

If *the gains* $\mathbf{L}_{1o}(\mathbf{x}_1)$ *and* $\mathbf{L}_{2o}(\mathbf{x}_1)$ *exist such that* (61) *is asymptotically stable, then* $\hat{\mathbf{x}}_2(t) \to \mathbf{x}_2(t)$ *and* $\hat{\mathbf{m}}(t) \to \mathbf{m}(t)$ *as* $t \to \infty$. ∎

Example 3: Chua's circuit [50]

Consider the Chua circuit described by (8), modified here with an additive message m as,

$$\begin{cases} \dot{\mathbf{x}} = \begin{bmatrix} \alpha(x_2 - x_1^3 - cx_1) + m \\ x_1 - x_2 + x_3 \\ -\beta x_2 \end{bmatrix} = \mathbf{f}(\mathbf{x}) + \begin{bmatrix} 1 \\ 0 \\ 0 \end{bmatrix} m \triangleq \mathbf{f}(\mathbf{x}) + \mathbf{B}_m(\mathbf{x})m, \\ \dot{m} = A_m m, \end{cases} \tag{62}$$

where $\alpha = 10$, $\beta = 16$ and $c = -0.14$ are the chaotic parameters [50]. A different modification scheme is given in Ref. [51].

Using (52), let the output be chosen as

$$\mathbf{y} = \begin{bmatrix} y_1 \\ y_2 \end{bmatrix} = \begin{bmatrix} x_1 \\ x_2 \end{bmatrix} \triangleq \mathbf{x}_1 \Rightarrow \mathbf{w} = \begin{bmatrix} x_3 \\ m \end{bmatrix}, \tag{63}$$

where x_3 constitutes the unknown partial state, and the derived measurement can be obtained as

$$\mathbf{y}_d = \underbrace{\begin{bmatrix} \dot{y}_1 \\ \dot{y}_2 \end{bmatrix}}_{\dot{\mathbf{y}}} - \underbrace{\begin{bmatrix} \alpha(x_2 - x_1^3 - cx_1) \\ x_1 - x_2 \end{bmatrix}}_{\mathbf{f}_w(y_1, y_2)} = \underbrace{\begin{bmatrix} 0 & 1 \\ 1 & 0 \end{bmatrix}}_{\mathbf{B}_w} \underbrace{\begin{bmatrix} x_3 \\ m \end{bmatrix}}_{\mathbf{w}}. \tag{64}$$

Using (62) and (63), the combined partial-state and message system has the form

$$\dot{\mathbf{w}} = \mathbf{A}_w \mathbf{w} + \mathbf{g}(y_2), \tag{65}$$

where $\mathbf{A}_w = \mathbf{0}$ if $A_m = 0$, $\mathbf{g}(y_2) = [-\beta x_2 \quad 0]^T$, and $[\mathbf{A}_w, \mathbf{B}_w]$ is an observable pair for all $\mathbf{A}_w$, i.e., $\text{rank}\left[\mathbf{B}_w^T \quad \mathbf{A}_w^T\mathbf{B}_w^T\right] = 2$.

Using (61) or (65) and (62), a UIO for partial-state and message estimations can be constructed based on $\mathbf{y}_d$ given by (64) and implemented as

$$\begin{cases} \dot{\mathbf{x}} = \mathbf{f}(\mathbf{x}) + \mathbf{B}_m(\mathbf{x})m, \quad \mathbf{x}(0) = \mathbf{x}_o, \\ \dot{\mathbf{z}} = (\mathbf{A}_w - \mathbf{L}_o\mathbf{B}_w)\mathbf{z} + \mathbf{g}(y_2) + \left[(\mathbf{A}_w - \mathbf{L}_o\mathbf{B}_w)\mathbf{L}_o\mathbf{y} - \mathbf{L}_o\mathbf{f}_w(y_1, y_2)\right], \quad \mathbf{z}(0) = \mathbf{z}_o, \\ \hat{\mathbf{w}} = \mathbf{z} + \mathbf{L}_o\mathbf{y}, \quad (\mathbf{A}_w = \mathbf{0}, \quad \mathbf{p}(\mathbf{x}_1) = \mathbf{L}_o\mathbf{y}, \quad \mathbf{y} = \mathbf{x}_1). \end{cases} \tag{66}$$

The gain $\mathbf{L}_o$ used for the simulations was obtained by choosing the UIO poles as $\mathbf{p}_o = -[\mathbf{1000} \ \mathbf{500}]$. Using Matlab's pole-placement command `Lo = place(Aw',Bw',po)'`, we obtain, with $\mathbf{A}_w = \mathbf{0}$,

Lo = 0	1000
500	0

The message $m(t)$ in (66) is a stock price data consisting of 50 data points where the value of $m(0)$ is $m(0) = 37$. To minimize the effect of $m(t)$ on the chaotic nature of the Chua circuit, it is scaled down to a small signal as $m(t) = 0.01m(t)$; this yields $\overline{m}(0) = 0.37$. The scaled down signal was then injected into (62) directly. To enhance the estimate $\mathbf{z}(t) = \hat{\mathbf{w}}(t) - \mathbf{L}_o\mathbf{y}(t) = \left[\hat{x}_3(t)\hat{\overline{m}}(t)\right]^T - \mathbf{L}_o[x_1(t) \quad x_2(t)]^T$, its initial value $\mathbf{z}(0)$ was calculated by using $\mathbf{z}(0) = \left[\hat{x}_3(0)\hat{\overline{m}}(0)\right]^T - \mathbf{L}_o[x_1(0) \quad x_2(0)]^T$, which gave $\mathbf{z}(0) = [498 \quad -999.63]^T$, where $\hat{x}_3(0) = x_3(0) = -2$ and $\hat{\overline{m}}(0) = \overline{m}(0) = 0.37$. We remark that, since the initial condition $\mathbf{x}(0) = [2 \quad -0.5 \quad -2]^T$ of the Chua circuit is known, we can always set $\hat{x}_3(0) = x_3(0)$, while in the event that the value of $\overline{m}(0)$ is not known, then it can be set as $\overline{m}(0) = 0$ resulting in small mismatches between $\hat{\overline{m}}(t)$ and $\overline{m}(t)$ during the transient period. The performance of the reduced-order UIO is shown in **Figures 6** and **7**. **Figure 6(a)** shows $x_3(t)$ and its estimate $\hat{x}_3(t)$, while **Figure 6(c)** shows the plot of $x_3(t)$ vs. $\hat{x}_3(t)$, which indicates an excellent match. **Figure 6(b)** displays the message $m(t)$ and its estimate $\hat{m}(t)$, while the plot of $m(t)$ vs. $\hat{m}(t)$ in **Figure 6(d)** shows a clean 45-degree line indicating an almost perfect match. The plots of $\{x_i(t), \ \overline{m}(t) \neq 0\}$ vs. $\{x_i(t), \ \overline{m}(t) = 0\}$ are depicted in **Figure 7(a)–(c)**, showing that the small signal $\overline{m}(t)$ has little effect on the chaotic nature of the Chua circuit. We conclude that the performance of the reduced-order UIO for partial-state and message estimations was satisfactory. Further, it is emphasized that no Jacobian linearization was employed in this example.

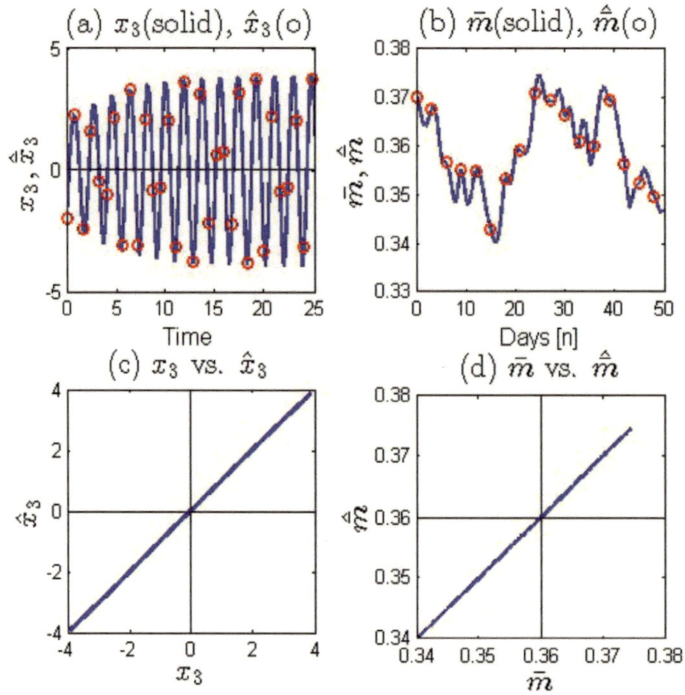

Figure 6. Responses of Chua system: (a) x_3 and $\hat{x}_3$; (b) $\bar{m}$ and $\hat{\bar{m}}$; (c) x_3 vs. $\hat{x}_3$; and (d) $\bar{m}$ vs. $\hat{\bar{m}}$. Figures 6(c) and 6(d) indicate negligible estimation errors $\tilde{x}_2 = x_2 - \hat{x}_2$ and $\tilde{\bar{m}} = \bar{m} - \hat{\bar{m}}$, respectively.

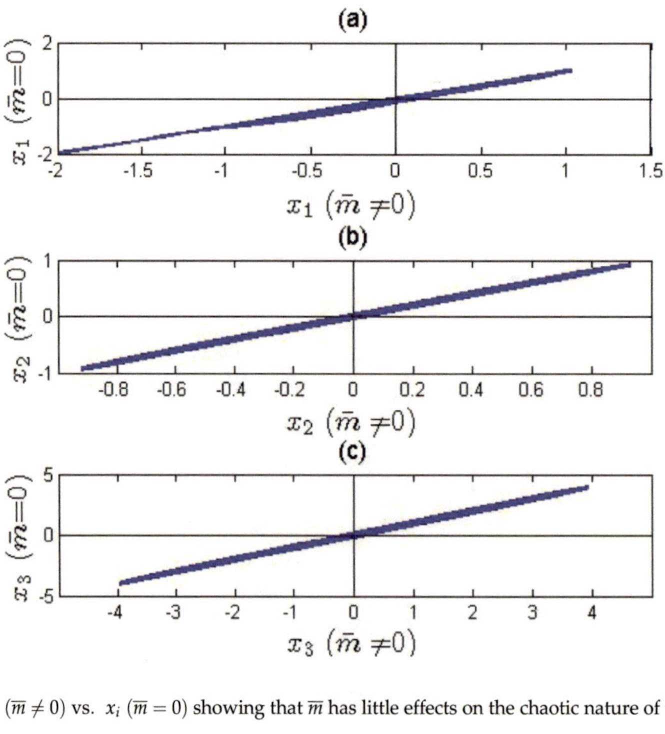

Figure 7. Plots of x_i $(\bar{m} \neq 0)$ vs. x_i $(\bar{m} = 0)$ showing that $\bar{m}$ has little effects on the chaotic nature of the Chua system.

5. Conclusions and plan for future research

In this paper, we showed that secure communication employing chaotic systems can be achieved by synchronizing the dynamics of the drive and response systems. The results are obtained by using unknown-input observers (UIOs), which serve as the response systems. Three classes of UIOs have been designed, namely, (i) full-order UIO for estimating all the state variables (full state) and messages in the drive system; (ii) reduced-order UIO for message estimation based on a derived measurement technique, where the formulation is based on the disturbance observer-based control (DOBC) theory (recall that the DOBC technique is only applicable to disturbance estimation based on the assumption that all the state variables (full state) in a system are known; and (iii) reduced-order UIO for partial-state and message estimations based on partial-state measurement using the derived-measurement technique. The reduced-order UIO for partial-state and message estimations is novel, and is an extension of the DOBC theory, thereby expanding the technique and applications of DOBC. Our future research and development will be focused on wireless secure communication, robust synchronization in the presence of channel noise and various channel induced distortions, and the designs and applications of disturbance cancellation nonlinear control systems using the well-known disturbance accommodation control (DAC) theory, thereby unifying the DAC and DOBC approaches and techniques.

Author details

Robert N.K. Loh* and Manohar K. Das

*Address all correspondence to: loh@oakland.edu or das@oakland.edu

School of Engineering and Computer Science, Oakland University, Rochester, Michigan, USA

References

[1] Rehan K, Qiao G. A survey of underwater acoustic communication and networking techniques. Research Journal of Applied Sciences, Engineering and Technology. 2013;**5**: 778–789

[2] Haug OT. Acoustic communication for use in underwater sensor networks [thesis]. Norwegian University of Science and Technology; 2009

[3] Zielinski A. Communications underwater, Invited presentation. Journal of Hydroacoustics. 2004;**7**:235–252

[4] Stojanovic M. Underwater acoustic communication. In: Wiley Encyclopedia of Electrical and Electronics Engineering. Hoboken, New Jersey, USA: Wiley Interscience; 1999

[5] Akyildiz IF, et al. Challenges for efficient communication in underwater acoustic sensor networks. CM SIGBED Review. Embedded Sensor Networks and Wireless Computing. 2004;**1**(Special issue):3–8

[6] Melodia T, et al. Advances in underwater acoustic networking. In: Mobile Ad Hoc Networking: Cutting Edge Directions. 2nd ed. Wiley Online Library. Chapter 23; 2013

[7] Yue J-P, et al. A robust wide synchronous digital chaotic communication scheme in shallow water channel. In: Proceedings of the Fourth International Workshop on Chaos-Fractals Theories and Applications (IWCFTA). 19–22 October 2011. New York: IEEE; 2011. pp. 203–207

[8] Shu X, et al. Chaotic modulations and performance analysis for digital underwater acoustic communications. Applied Acoustics. 2016;**105**:200–208

[9] Shu X, et al. Chaotic direct sequence spread spectrum for secure underwater acoustic communication. Applied Acoustics. 2016:**104**:57–66

[10] Ren H-P, et al. A chaotic spread spectrum system for underwater acoustic communication. Physica A: Statistical Mechanics and its Applications. 2017;**478**:77–92

[11] Shu X, et al. Underwater chaos-based DS-CDMA system. In: Proceedings of IEEE International Conference on Signal Processing Communications and Computing. 19–22 Sept. 2015. New York: IEEE; 2015

[12] Azou S, Burel G, Pistre C. A chaotic direct-sequence spread-spectrum system for underwater communication. In: Proceedings of IEEE Oceans. 29–31 Oct. 2002. New York: IEEE; 2002. pp. 2409–2415

[13] Ren H-P, et al. Experimental validation of wireless communication with chaos. Chaos. 2016;**26**:083117

[14] Ren H-P, et al. Wireless communication with chaos. Physics Review Letters. 2013;**110**: 184101

[15] Yao J-L, et al. Chaos-based wireless communication resisting multipath effects. arXiv: 1611.02325 [cs.IT]

[16] Kaddoum G. Wireless chaos-based communication systems: A comprehensive survey. IEEE Access. 2016;**4**:2621–2648

[17] Al-Hussaibi, Walid A. Effect of filtering on the synchronization and performance of chaos-based secure communication over Rayleigh fading channel. Communications in Nonlinear Science and Numerical Simulation. 2015;**26**:87–97

[18] Al-Hussaibi, Walid A, et al. On the synchronization of secure wireless communication systems based on chaotic circuits. Open Transactions on Wireless Communications. In Press. Available from: www.scipublish.com/journals/WSN/papers/download/3306–1428.pdf [Accessed: 15-06-2017].

[19] Jiancheng Z, Fanglai Z. Chaos synchronization and chaos-based secure communication based on new unknown input observer approach. In: Proceedings of the 35th Chinese Control Conference. 27–29 July 2016. New York: IEEE; 2016. pp. 1891–1896

[20] Yang J, Zhu F. Synchronization for chaotic systems and chaos-based secure communications via both reduced-order and step-by-step sliding mode observers. Communications in Nonlinear Science and Numerical Simulation. 2013;**18**:926–937

[21] Yang J, Chen Y, Zhu F. Associated observer-based synchronization for uncertain chaotic systems subject to channel noise and chaos-based secure communication. Neurocomputing Archive. 2015;**167**:587–595

[22] Yang J, Zhu F. Synchronization for chaos-based secure communications via reduced-order and step-by-step sliding mode observers. Communications in Nonlinear Science and Numerical Simulation. 2013;**18**:926–937

[23] Chang W-D, Shih S-P, Chen C-Y. Chaotic secure communication systems with an adaptive state observer, Journal of Control Science and Engineering. 2015;**2015**:1–7. Available from: http://dx.doi.org/10.1155/2015/471913 [Accessed: 28-03-2017]

[24] Johnson CD. Accommodation of disturbances in linear regulators and servomechanism problems. IEEE Transactions on Automatic Control. 1971;**AC-16**:635–644

[25] Johnson CD. On observers with unknown and inaccessible inputs. International Journal of Control. 1975;**21**:825–831

[26] Johnson CD. Theory of disturbance-accommodating controllers. In: Control and Dynamic Systems, Advances in Theory and Applications. In: Leondes CT, editor. Cambridge, MA, USA: Academic Press; 1976. pp. 387–489

[27] Ciccarella G, Dalla Mora M, Germani A. A Luenberger-like observer for nonlinear systems. International Journal of Control. 1993;**57**:537–556

[28] Mora MD, Germani A, Manes C. A state observer for nonlinear dynamical systems. Nonlinear Analysis, Theory, Methods, & Applications. 1997;**30**:4485–4496

[29] Friedland B. Advanced Control System Design. New Jersey, USA: Prentice-Hall; 1996

[30] Glad T, Ljung L. Control Theory, Multivariable and Nonlinear Methods. Abingdon, Oxfordshire, UK: Taylor and Francis. 2000

[31] Besancom G. An Overview on Observer Tools for Nonlinear Systems. In: Beancon G, Editor. Nonlinear Observers and Applications. Berlin, Germany: Springer-Verlag. 2007

[32] Kravaris C, Sotiropoulos V, Georgiou C, Kazantzis N, Xiao MQ, and Krener AJ. Nonlinear observer design for state and disturbance estimation. Systems & Control Letters. 2007;**56**:730–735

[33] Sundarapandian V. Reduced-order observer design for nonlinear systems. Applied Mathematics Letters.; 2006;**19**:936–941

[34] Trinh H, Fernando T. Functional Observers for Dynamical Systems. Berlin, Germany: Springer-Verlag. 2011. p. 227

[35] Pecora LM, Carroll T. Synchronization in chaotic systems. Physical Review Letters. 1990;**64**:821–824

[36] Inoue E, Ushio T. Chaos communication using unknown input observers. Electronics and Communication in Japan. 2001;**84**(12):21–27

[37] Feki M. An adaptive chaos synchronization scheme applied to secure communication. Chaos, Solitons and Fractals. 2003;**18**:141–148

[38] Chen M, Zhou DG, Shang Y. A new-observer-based synchronization scheme for private communication. Chaos, Solitons and Fractals. 2005;**24**:1025–1030

[39] Chen M, Wang M. Unknown input observer based chaotic secure communication. Physics Letters A. 2008;**372**:1595–1600

[40] Wang XY, Wang MJ. A chaotic communication scheme based on observer. Communication in Nonlinear Science and Numerical Simulation. 2009;**14**:1502–1508

[41] Grzybowski JMV, Eisencraft M, Macau EEN. Chaos-based communication systems: Current trends and challenges. In: Banerjee S, et al., editors. Application and Nonlinear Dynamics in Engineering; 1: Understanding Complex Systems. Heidelberg: Springer-Verlag; 2011. Chapter 7

[42] Mata-Machuca JL, Martinez-Guerra R, Aguilar-Lopez R. A chaotic system in synchronization and secure communications. Communications in Nonlinear Science and Numerical Simulation. 2012;**17**:1706–1713

[43] Yau HT, Pu YC, Li S. Application of a chaotic synchronization system to secure communication. Information Technology and Control. 2012;**41**:274–282

[44] Dimassi H, Loria A, Belghth S. A new secured transmission scheme based on chaotic synchronization via smooth adaptive unknown input observers. Communications in Nonlinear Science and Numerical Simulation. 2012;**17**:3727–3739

[45] Wang XY, Wang MJ. A chaotic secure communication scheme based on observer. Communications in Nonlinear Science and Numerical Simulations. 2009;**14**:1502–1508

[46] Mata-Machuca J, Martinez-Guerra R, Aguilar-Lopez R. A chaotic system in synchronization and secure communications. Communications in Nonlinear Science and Numerical Simulation. 2012;**17**:1706–1713

[47] Rossler OE. An equation for continuous chaos. Physics Letters A. 1976;**57**:397–398

[48] Strogatz ST. Nonlinear Dynamics and Chaos with Applications to Physics, Biology, Chemistry, and Engineering. 2nd ed. Boulder: Westview Press; 2015

[49] Genesio R, Tesi A. A harmonic balance methods for the analysis of chaotic dynamics in nonlinear systems. Automatica. 1992;**28**:531–548

[50] Chua L, Komuro M, Matsumoto T. The double scroll family. IEEE Transactions on Circuits and Systems. 1986;**33**:1072–1178

[51] Mauricio Zapateiro De la Hoz M, Acho L, Vidal Y. A secure communication system based on a modified chaotic Chua oscillator. In: Proceedings of 7th CHAOS Conference. 7–10 June 2014; Lisbon. Oakland: ISAST. pp. 563–572

[52] Moreno J. Unknown input observers for SISO nonlinear systems. In: Proceedings of 39th IEEE CDC Conference. 12–15 December 2000; Sydney. New York: IEEE, 2000. pp. 790–795

[53] Isidori A. Nonlinear Control Systems. Springer; 1995: p. 549

[54] Slotine JJ, Li WP. Applied Nonlinear Control. New Jersey: Prentice Hall; 1991: p. 459

[55] Kalman RE, Bucy RS. New results in linear filtering and prediction theory. ASME Journal of Basic Engineering. 1961;**83**:95–107

[56] Gelb A. Applied Optimal Estimation. Massachusetts: MIT Press; 1974: p. 374

[57] Luenberger DG. An introduction to observers. IEEE Transactions on Automatic Control. 1971;**AC-16**:596–602

[58] Gopinath B. On the control of linear multiple input-output systems. American telephone and telegraph company—The Bell System Technical Journal. 1971;**50**:1063–1081

[59] Chen CT. Linear System Theory and Design. 4th ed. New York: Oxford University Press; 2013: p. 400

[60] Ogata K. Modern Control Theory. 5th ed. New Jersey: Prentice Hall; 2010: p. 894

[61] Li S, Jun Y, Chen W-H, Chen X. Disturbance observer-based control. Methods and Applications. Boca Raton: CRC Press; 2014: p. 342

Chapter 4

Denoising Methods for Underwater Acoustic Signal

Vijaya Baskar Veeraiyan

Additional information is available at the end of the chapter

http://dx.doi.org/10.5772/intechopen.69027

Abstract

Underwater ambient noise is primarily a background noise which is a function of time, location, and depth. It is of prime importance to detect the signals such as sound of a submarine or echo from a target surpassing this ambient noise. It is also defined as the residual noise that remains after all easily identifiable sound sources are eliminated. In the absence of the sound from ships and marine life, underwater ambient noise levels are dependent mainly on wind speeds at frequencies between 500 Hz and 50 KHz.The detection of background noise is essential to enhance the signal-to-noise ratio of acoustic-based underwater instruments. Since there is a possibility of signal and noise present in the same frequency, it becomes essential to find out a suitable algorithm to perform denoising. In this chapter, various denoising techniques such as wavelet, empirical mode decomposition (EMD) in time domain, ensemble empirical mode decomposition (EEMD), and frequency domain-based EMD are studied, and the results are compared. The proposed frequency domain algorithm produced better results in the frequency ranging from 50 Hz to 25 KHz, with less signal error.

Keywords: underwater acoustics, ambient noise, denoising, wavelet, EMD

1. Introduction

The ocean is filled with sound. Underwater sound is generated by a variety of natural sources, such as breaking waves, rain, and marine life [1, 2]. It is also generated by a variety of man-made sources, such as ships and military sonars. Some sounds are present more or less everywhere in the ocean all of the time. The background sound in the ocean is called ambient noise [3, 4]. Ambient noise also excludes all forms of self-noise, such as the noise of current flow around the measurement hydrophone and its supporting structure and obviously must exclude all forms of electrical noise (Urick). In shallow water (depth of 10–100 m), acoustic systems like sonar suffer a huge loss of acoustic signals due to ambient noise [5]. Wind noise is the most constant and most perennial of all components of ambient noise [3, 6]. Many theoreticians have predicted ambient noise that is caused by wind. Different actions are prevailing in a dissimilar component of the entire frequency band from 1 to 50 KHz [2]. The recovery of the signal buried in ambient noise is of important significance for target's signal detection, recognition, and classification at low signal-to-noise ratio.

2. Denoising methods

In science and engineering, noise is defined as an unwanted signal, particularly a stochastic and persistent form that dilutes the lucidity of a signal. In natural process, noise could be caused by the process itself, such as local and intermittent instabilities, irresolvable subgrid phenomena, or some concurrent processes in the environment in which the investigations are conducted. It could also be generated by the sensors and recording systems when observations are made. In general, all data are amalgamations of signal and noise, i.e.,

$$x(t) = s(t) + n(t) \tag{1}$$

in which $x(t)$ is the recorded data, and $s(t)$ and $n(t)$ are the true signal and noise, respectively.

The general algorithm of denoising can be written as:

i. Convert the data from a time domain into a suitable domain where signal and noise can be separated.

ii. Remove noise in the new domain using suitable algorithm.

iii. Then convert the data back to the time domain.

Identifying a suitable domain to eliminate the noise completely is a big challenge. Denoising has been performed using different denoising methods [wavelet, empirical mode decomposition (EMD), and ensemble EMD (EEMD)]. Here, a simulated signal was considered to be the sonar signal, and it was added with a real-time wind noise signal to get a noisy signal. This signal was applied as the stimulus to the denoising algorithm, and the denoised signal was obtained.

3. Denoising using wavelets

The wavelet transform is extensively used in several fields of processing a signal. It has the edge of employing variable sized time window for dissimilar bands of frequency. It has an advantage of higher resolution in frequency, particularly in lower bands and low resolution, when it comes to higher frequency bands. For the modeling of nonstationary signals, for example, speech that possesses dull temporal variations when it comes to low frequency and sudden temporal deflections in higher frequency, the wavelet transform is a perfect answer. This is also used for denoising of the signal as well.

Denoising of underwater acoustic using wavelet is engineered in general with the following steps in place:

1. Source signal to be taken.
2. Real-time wind noise data are added to the source signal in obtaining the noisy signal.
3. Wavelet decomposition is carried out for the noisy signal using a suitable wavelet.
4. Updating noisy coefficients of a level should be done by setting threshold values. This setting is done by considering detailed noise coefficients.
5. In a similar manner, to update the approximate noisy coefficients of a level, the last level approximation noise coefficient is used to set the threshold value.
6. Next step is to compute modified noisy coefficient using different threshold functions such as hard, soft, and nonnegative garrote.
7. Signal retrieval or denoised signal is done by applying inverse wavelet transform.
8. Calculate the error in the denoised signal for different threshold functions and also for different wavelets.

Here, a universal method for fixing the threshold value is modified by introducing two constants 'k' and '$m1$' to obtain higher quality output signal, and it is combined with nonnegative garrote threshold function in the denoising process.

The modified threshold equation is given by Aggarwal et al. [7].

$$\lambda = k.m1.\sigma\sqrt{2\log2(N)} \tag{2}$$

Where N denotes the number of samples of noise. It is noted that if two factors, i.e., k and $m1$ are introduced in the universal threshold equation, then new threshold value gives better results, especially to recover the original signal [1].

Here, the values of k and $m1$ are fixed after repeated trials. Initially, $m1$ value was fixed, and k value varied to obtain better result. After many trials, k value is fixed as '0.5', which gives a better result when compared to other k values. The root-mean-square error (RMSE) value is low when $k = 0.5$, and there is no change in the RMSE value above this k value. So k value was fixed as 0.5. Then, the value of $m1$ is varied, and the output was better for the value of $m1 = 3$. The RMSE value is less for $m1 = 3$, and above this value, there is no change in the RMSE value. So it has been concluded that the value of k is fixed as $k = 0.5$ and $m1 = 3$ to get better results.

The final step is by enforcing inverse wavelet transform to obtain the original signal. The denoised signal thus obtained should be similar to the original signal, i.e., RMSE of the signal should be low.

3.1. Denoising results obtained using wavelets

In this work, sine wave is taken as a sample signal. This signal is added with a real-time wind noise signal. This noisy signal is processed with the denoising steps, i.e., decomposition, wavelet thresholding, and inverse wavelet transform for reconstruction. Initially, the signal decomposition is done for a different level. After a series of test, it was found that the decomposition level 2 gives better output. Now for each level, the threshold values are calculated using the modified universal threshold equation. Then thresholding is applied to detailed noisy coefficient in order to obtain the modified detailed coefficient. The denoised signal was reconstructed using the true approximation coefficient of level 2 and the modified detailed coefficient of levels 1– 2.

In this work, decomposition was done using different wavelets. After decomposition by wavelet, the threshold is applied using different threshold functions, i.e., soft, hard, and nonnegative garrote threshold [8]. The RMSE value was calculated for different wavelet and threshold function. These values are presented in **Table 1**. From the table, it is clear that RMSE value is less for 'sym8' wavelet compared with other wavelets and most of the wavelets perform well along with nonnegative garrote threshold, i.e., RMSE value is less.

Then, the modified universal threshold along with nonnegative garrote threshold is applied to the decomposed signal. After applying threshold value, inverse wavelet is applied at last to obtain the original signal. The signal, noise, noisy, and the denoised signals are shown in **Figure 1**. The qualitative output, i.e., the comparison of original and denoised signals is presented in **Figure 2**.

Wavelet type	RMSE value for different threshold method		
	Hard	Soft	Nonnegative garrote
haar	0.001339103	0.001303089	0.001326211
db2	0.000836217	0.000837207	0.000836016
db4	0.000618673	0.000620417	0.000618689
db5	0.000602552	0.000603077	0.000602235
db8	0.000595174	0.000595174	0.000595174
sym4	0.000617997	0.000619197	0.000617785
sym8	0.000594376	0.000595753	0.000594325
coif4	0.000594698	0.000596959	0.000594531
coif2	0.000615408	0.000617598	0.000615114
dmey	0.000593861	0.000593861	0.000593861

Table 1. RMSE value for different threshold function.

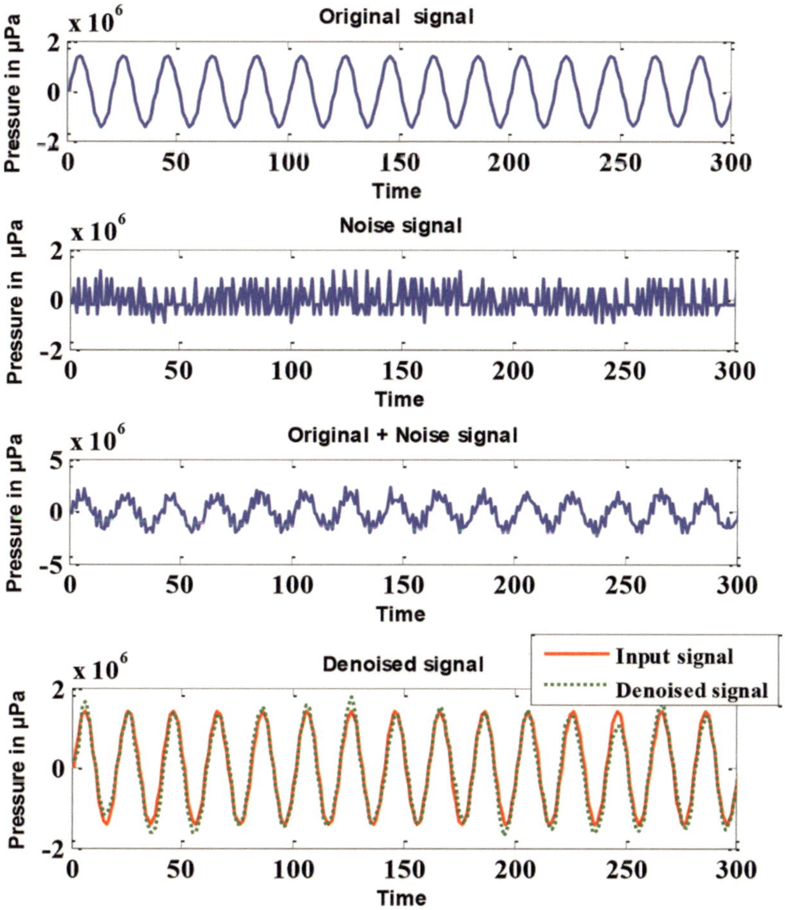

Figure 1. Process of denoising using wavelet.

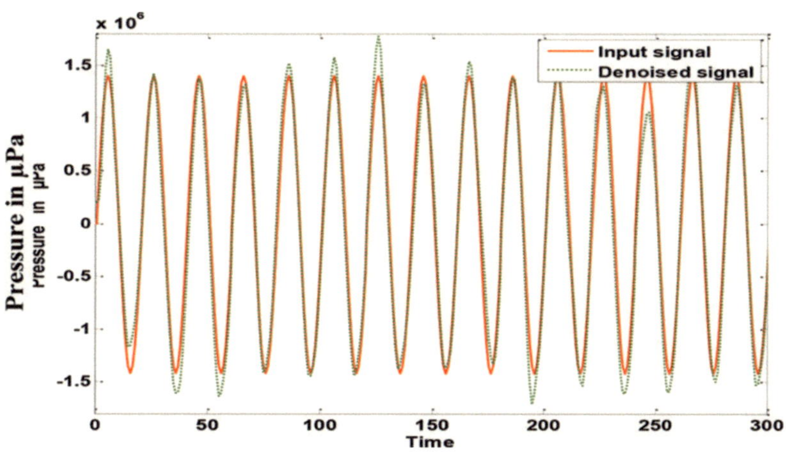

Figure 2. Input and denoised output signal comparison graph.

4. Denoising using EMD

Empirical mode decomposition (EMD) is different from other methods of analyzing data from nonstationary and nonlinear processes. This has been introduced by Huang et al.. This is used to decompose signals in an adaptive manner into a sum of AM and FM components containing raw intrinsic building blocks. These define the complex waveform. There is no need to fix the functional basis in advance. EMD sifting methods are usually employed to procure these basis functions in an adaptive manner [9].

EMD identifies the intrinsic oscillatory modes by their characteristic time scales in the data through empirical observation and then decomposes the data into the corresponding IMFs through shifting processes. Intrinsic Mode Functions (IMF) may contain variable amplitude and frequency functions those are dependents of time. It is an algorithm to assign an instantaneous frequency to each IMF in order to decompose an arbitrary data set.

Since in EMD, decomposition is based on and derived from the data, it is an adaptive method. Here, the data $x(t)$ is decomposed in terms of IMFs, i.e.,

$$x(t) = \sum_{j=1}^{n} c_j(t) + r_n(t) \tag{3}$$

Here r_n is the residue of data $x(t)$, after n number of IMFs being extracted. IMFs are simple oscillatory functions with varying amplitudes and frequencies [10].

4.1. Denoising using EMD based on existing time domain approach

The importance of a given model is estimated with the information pertained to IMF datum at times where the noise makes its presence.

In this algorithm, EMD was applied to the noisy signal. Then, the noisy signal was decomposed into a set of IMFs. The energy of each IMF had been calculated and then the threshold value. The IMFs were shrunken using the nonnegative threshold function and then added to get the denoised output [1, 11, 12].

The energy of the first IMF is defined as

$$W_H[1] = \sum_{n=1}^{N} c_1^2(n) \tag{4}$$

c_1 represents the first IMF coefficients.

The energy of each IMF is defined as

$$\text{energy}(\text{count}) = \frac{\text{energy}(1)}{0.719} * 2.01^{-\text{count}}, \text{count} = 2, 3, \ldots \tag{5}$$

Where *energy* (1) is the noise energy that can be achieved by the first IMF variance using equation (4) and *count* is a variable that specifies the IMF number.

Then, the threshold value of each IMF can be calculated by

$$\text{threshold(count)} = \sqrt{\left(\frac{\text{energy(count)}}{x\text{size}}\right) * 2 * \log(x\text{size})} \tag{6}$$

The first IMF can be discarded as it captures most of the noise, and for other IMFs, the adaptive threshold can be calculated using Eq. (6). Then, the coefficients of each IMF was shrunken using nonnegative garrote threshold function that is calculated for values of IMF greater than or equal to the threshold value using

$$\text{mode (count)} = \text{mode(count)} - \left(\frac{\text{threshold (count)}^{\text{count}}}{\text{mode (count}^{(\text{count}-1)}}\right) \tag{7}$$

Where *mode* represents the IMF, and the variable count specifies the IMF number.

For values of IMF less than the threshold value,

$$\text{mode (count)} = 0 \tag{8}$$

Finally, the shrunken IMFs were added to obtain the denoised signal. The denoising algorithm is shown in **Figure 3**.

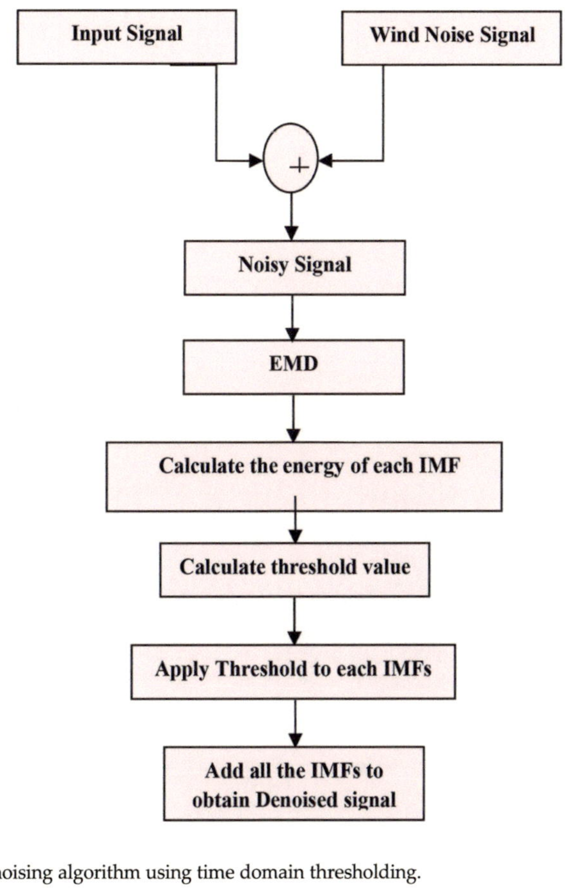

Figure 3. EMD-based denoising algorithm using time domain thresholding.

A sinusoidal signal was considered as the input test signal, and then, a real-time wind-driven ambient noise signal that was collected at the wind speed of 5.06 m/s was added with the input signal to obtain the noisy signal. The input, noise, and noisy signals are presented in **Figures 4–6**. The noisy signal was decomposed into a set of IMFs by using EMD function and is shown in **Figure 7**. Each of the IMF was shrunken using the nonnegative garrote threshold function. Then, the denoised signal was reconstructed by adding the shrunken IMFs, which is shown in **Figure 8**.

This algorithm performs well until the noise amplitude is less than the signal amplitude. The output is not acceptable whenever the noise amplitude is higher than 50% of the signal

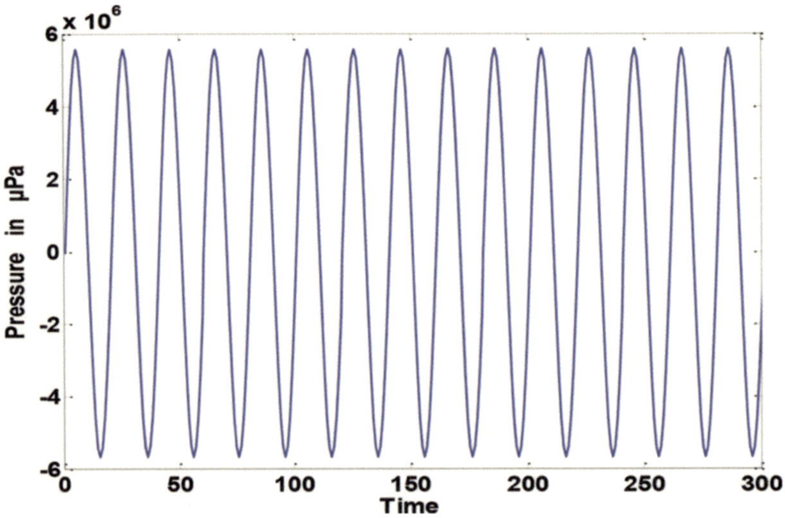

Figure 4. Input signal.

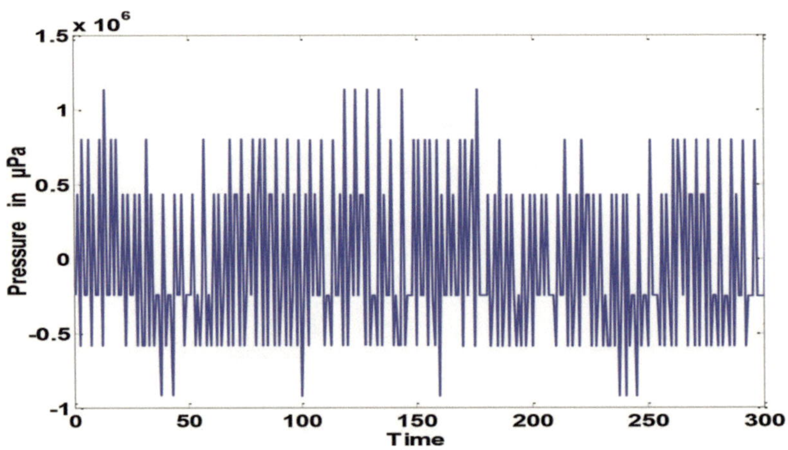

Figure 5. Wind noise signal.

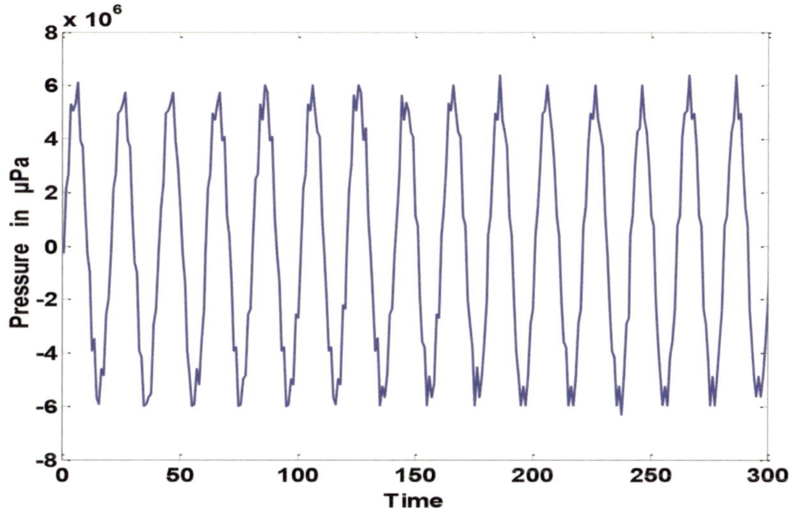

Figure 6. Noisy signal at 5.06 m/s.

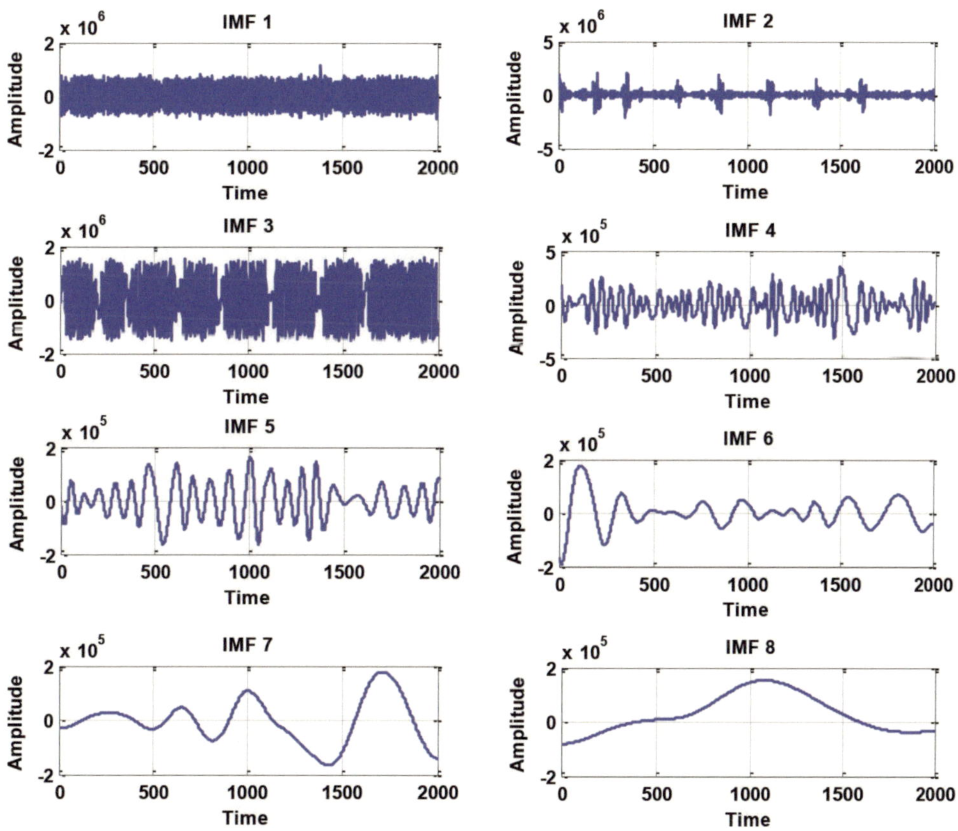

Figure 7. IMFs obtained from the noisy signal.

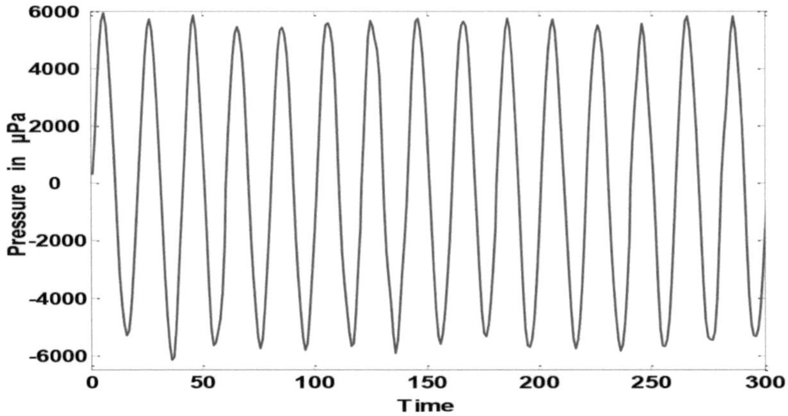

Figure 8. Denoised output signal for Time domain EMD.

amplitude. It is apparent from **Figures 4** and **8** that the denoised signal amplitude is lesser than the actual input signal; even though the denoised signal is similar to the input signal, it is not an exact resemblance to the input [1].

5. Denoising using EEMD

The output is not satisfactory in the EMD-based time domain method. So to improve the performance, the same time domain thresholding is used along with ensemble empirical mode decomposition (EEMD).

Ensemble EMD is a novel noise-assisted method of data analysis that has been proposed to surpass the scale separation problem without the subjective intermittence test. According to this, the true IMF components are defined as the mean of the ensemble of trials. Here, every IMF component comprises the signal along with white noise, whose amplitude is finite. It is very much feasible to isolate the scale without any selection of advanced subjective criterion in the ensemble approach. This novel method is reaped from the recent study of statistical properties of white noise [12] (Wu and Huang, 2004). This exemplifies the fact that EMD is an efficient adaptive dyadic filter bank when it is applied to white noise. Particularly, this novel method has the inspiration from the noise added analysis started by Flandrin et al (2005) and Gledhill (2003). The results of these experiments made it clear that noise is a useful tool in the analysis of data in the empirical mode decomposition [12].

The main feature of EEMD is the addition of white noise results in populating the whole time-frequency space equally, while the constituting components of dissimilar scales are isolated by the filter bank. In this state, the bits of the signal belongs to dissimilar scales would automatically proposed onto proper scales of reference given by the white noise as the background. Mode mixing can be avoided by the addition of finite noise, and hence, EEMD has an edge over the EMD method.

EEMD method is as follows:

1. Initially adding a series of white noise to the signal that is considered as a target.
2. Next is extracting IMF through decomposition of noise added data.
3. Repeating steps 1 and 2 several times, with dissimilar series of white noise every time.
4. Final results are the extraction of ensemble of corresponding IMFs of the decomposition [1].

As a matter of fact, the amplitude of the noise that is added is tiny, and it does not necessarily introduce the change of extreme upon which the EMD is dependent. The effect of white noise becomes negligible with the increase in ensemble members.

5.1. Denoising using EEMD based on time domain approach

The denoising process using EEMD has the same input signal and noise signal, which is used in Section 4.1, and is used to get the noisy signal. The noisy signal was applied as an input to the EEMD-based denoising algorithm. The input and output of EEMD algorithm are shown in **Figure 9**. In EEMD-based denoising algorithm, the output amplitude is the same as that of the

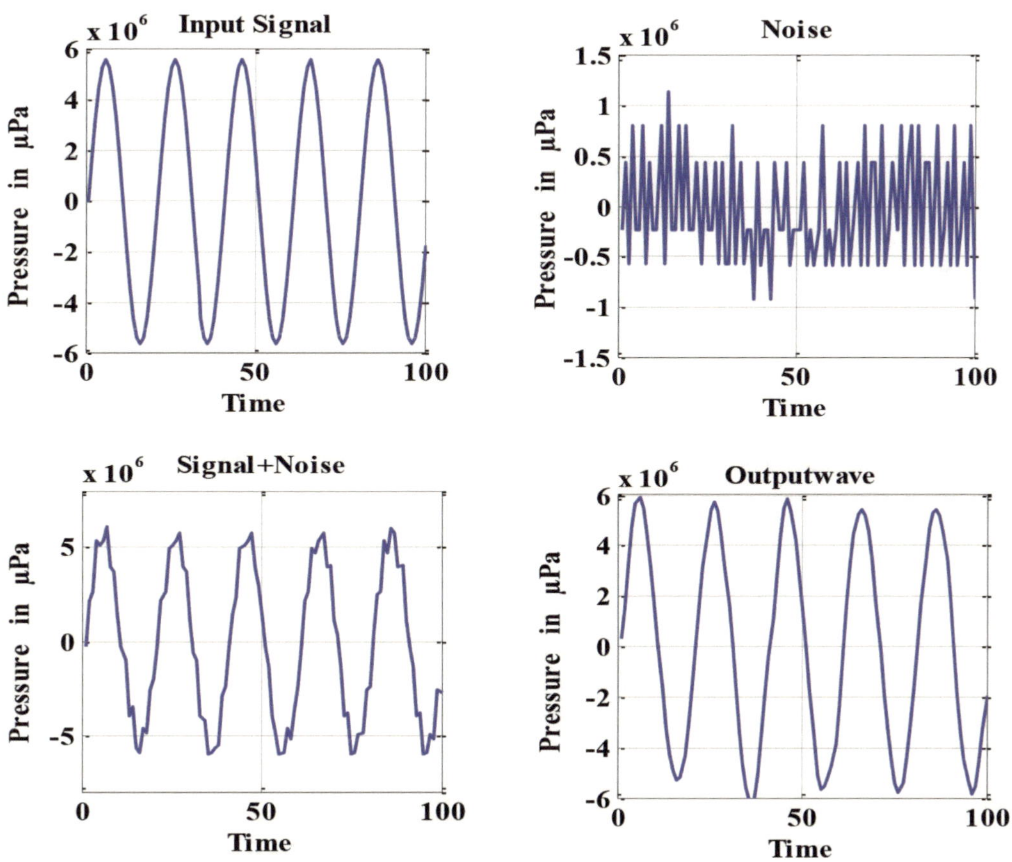

Figure 9. Input and output signals of EEMD algorithm.

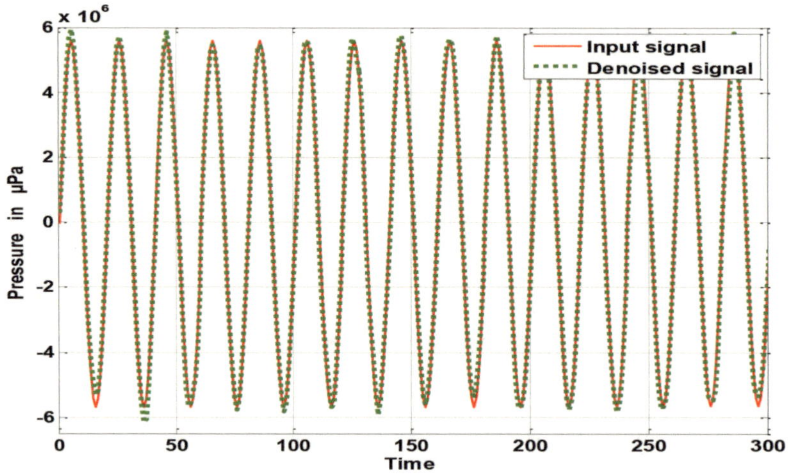

Figure 10. Input and denoised output signals comparison graph (EEMD method).

input signal amplitude, when compared to EMD-based denoising algorithm output. Even though the denoised signal amplitude is quite satisfactory, the output signal does not resemble the input signal exactly.

Comparison, between the input and denoised signal, which is obtained by EEMD, based time domain denoising algorithm, is presented in **Figure 10**. This algorithm produces better output compared to the output of EMD algorithm. But this algorithm takes more time to produce output, and also the output signal needs improvement.

6. Proposed denoising method using EMD based on frequency domain approach

In this proposed algorithm, EMD is used as a denoising tool. Earlier EMD-based denoising methods had been done using time domain thresholding.

The proposed algorithm is based on frequency domain thresholding, and the algorithm is shown in **Figure 11**. The proposed algorithm is simple, and it capable of producing better results than the existing algorithms. As the value of threshold depends on the noise signal, this algorithm performs well for different wind noise signals, i.e., noise collected for various wind speeds. From the results, it is concluded that the algorithm works well even if the signal amplitude and the noise amplitude are same. The signal shown in **Figure 12** is considered as an input signal. **Figure 13** shows the real-time wind-driven underwater ambient noise signal, which is measured at the wind speed of 5.06 m/s. The input signal is added with the noise signal.

The noisy signal is shown in **Figure 14**, which is noisier compared to the signal which are applied in the previous algorithms.

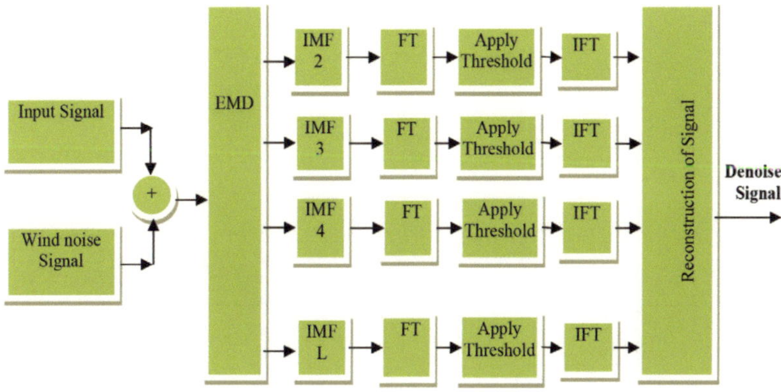

Figure 11. Proposed denoising algorithm based on EMD.

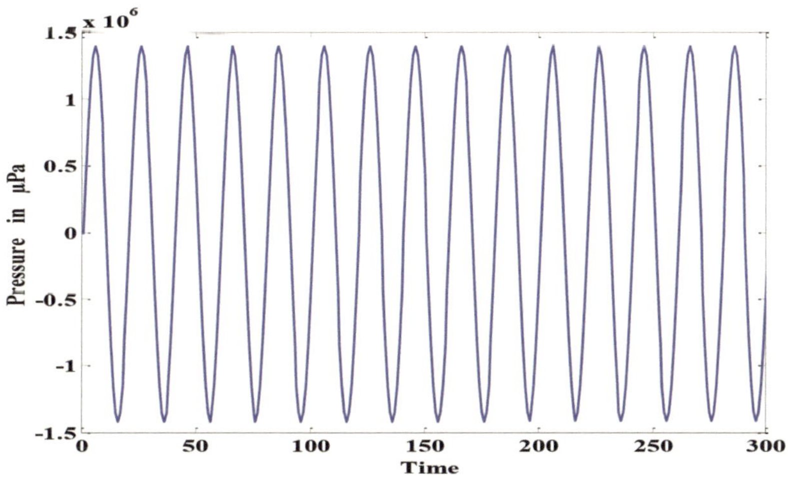

Figure 12. Input signal (0.0053 = 5 mv).

The noisy signal is decomposed into a set of IMFs by using EMD function. The IMFs are shown in **Figure 15**.

From the figure, it is clear that the IMF1 contains more noise. So we have eliminated IMF1. Then, we have applied FFT to other IMFs, which is shown in **Figure 16**.

From the figure, it is clear that IMF 2–4 have more noise components and IMF 5–8 have more signal components. Different threshold values have been used and better outputs are obtained, when the threshold value is set as 70–90% of maximum FFT amplitude, i.e., in each IMF, the coefficients of IMF signal, which have FFT amplitude below the threshold value, were assigned zero. After applying threshold, IFFT was taken to each IMF, and then, all the thresholded IMFs were added to get the denoised signal. The denoised signal is shown in **Figure 17** for different threshold values. The output was good for the threshold values of 70% and above. At 90% threshold, the output resembles the input signal very well. Compared to the existing time

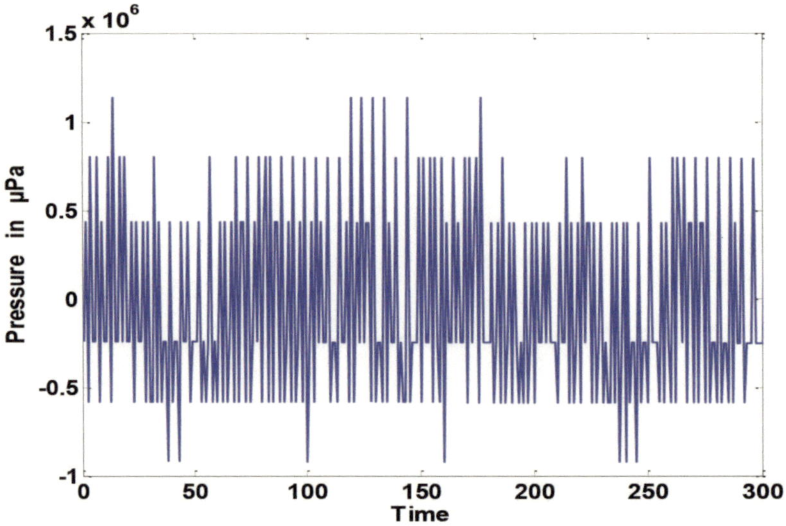

Figure 13. Noise signal.

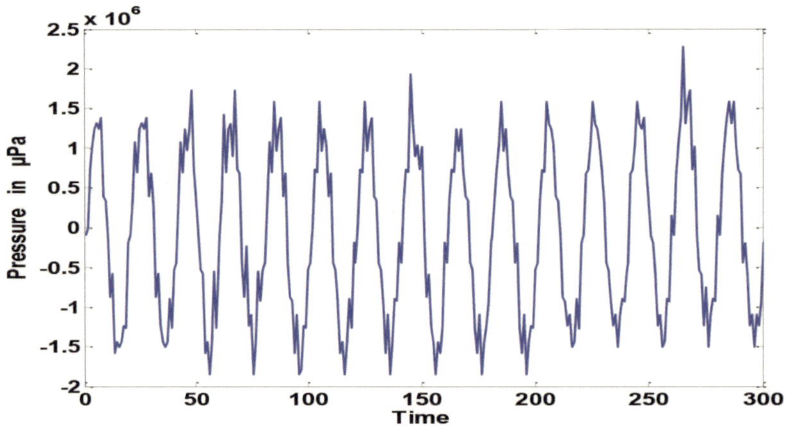

Figure 14. Noisy signal.

domain thresholding algorithm, the proposed frequency domain thresholding algorithm fetches better results.

In the existing algorithm, the signal amplitude is considered to be 20 mv, and in the proposed algorithm, the signal amplitude is considered to be 5 mv. In the existing algorithm, the denoised signal amplitude is much lesser than the actual input signal, and also, the output does not exactly resemble the input signal. But in the proposed algorithm, the amplitude of denoised signal (at 90% threshold) is same as that of the input signal, which is shown in **Figure 18**. Also, here, the denoised signal resemblance of the input signal is good. In all the results, the signal amplitude is represented in micropascal, i.e., volt is converted into

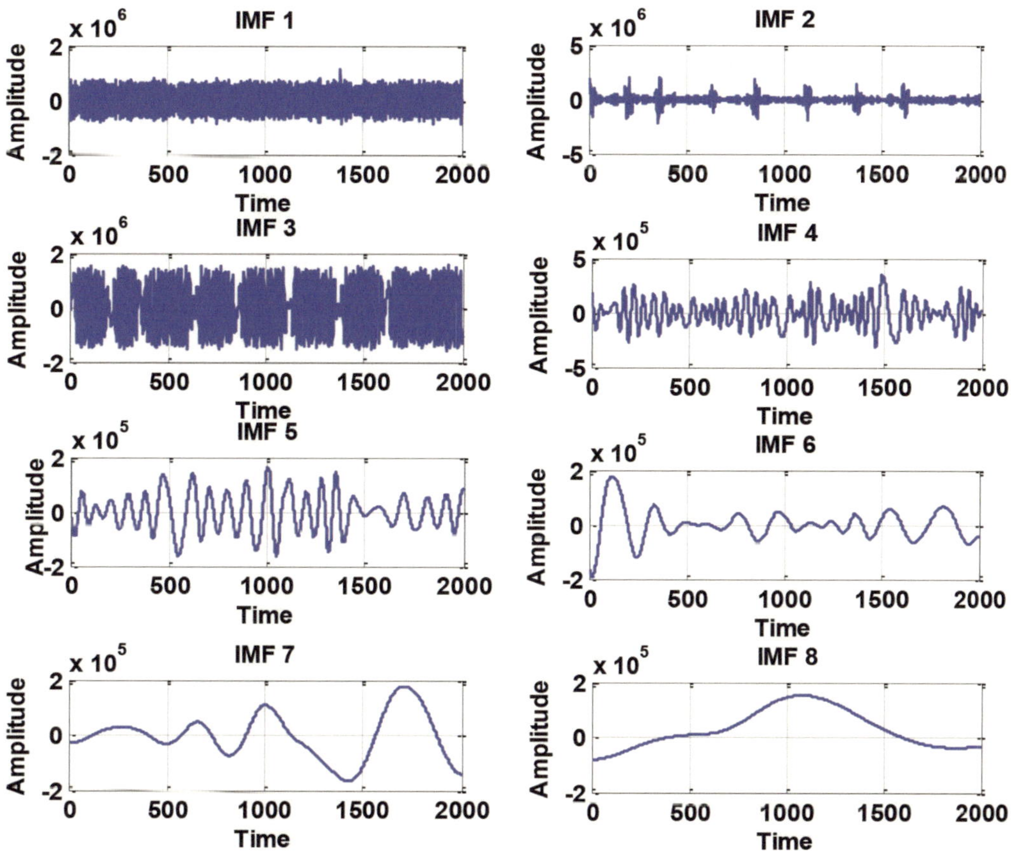

Figure 15. IMFs of noisy signal.

micropascal-based on the sensitivity of the microphone. This result again reveals the reliability of the algorithm for different wind noise signals.

This algorithm is tested even for chirp input signal. **Figure 19** shows the chirp input signal, noise, and noisy output signal of the proposed algorithm. **Figure 20** presents the IMFs of the noisy signal, which is produced by adding the chirp input with the wind noise signal (5.06 m/s). This IMF is different from one, which is shown in **Figure 15**. The comparison between the input and denoised signal is presented in **Figure 21**. From this figure, it is clear that the proposed algorithm performs well even for chirp input signal.

In order to validate the algorithm, the mean square error (MSE) is calculated using the following equation.

$$\text{Mean square error (MSE)} = \frac{1}{N} \sum_{n=1}^{N} (Z(n) - \hat{Z}(n))^2 \tag{9}$$

where N is the length of data, $Z(n)$ is the actual input signal, and $\hat{Z}(n)$ is the denoised signal.

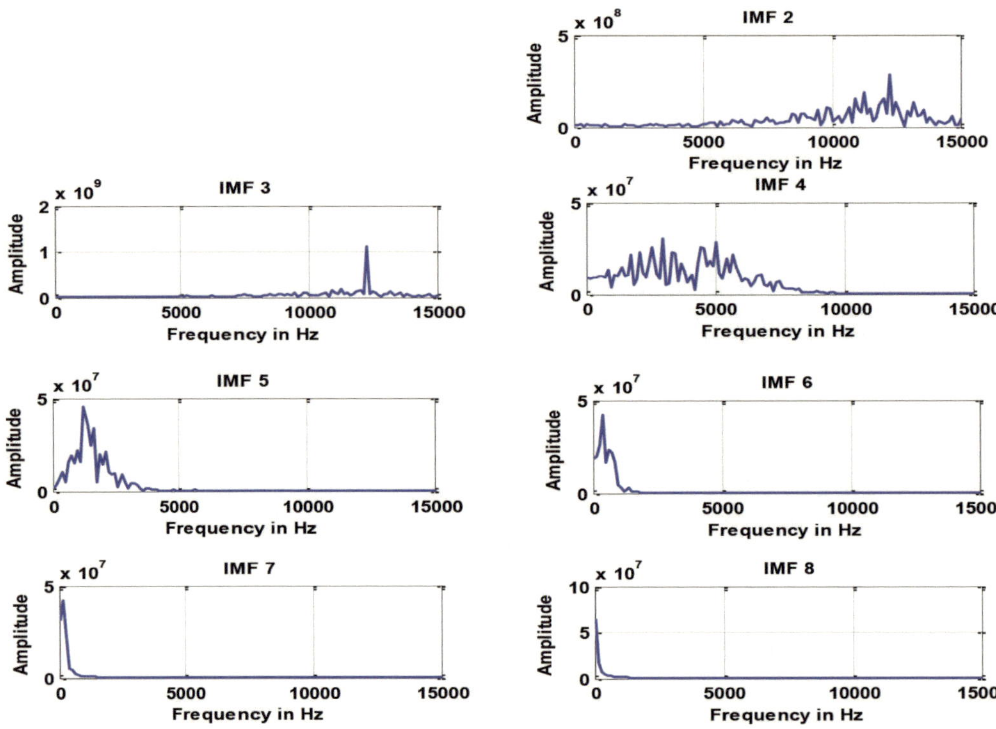

Figure 16. FFT of IMFs.

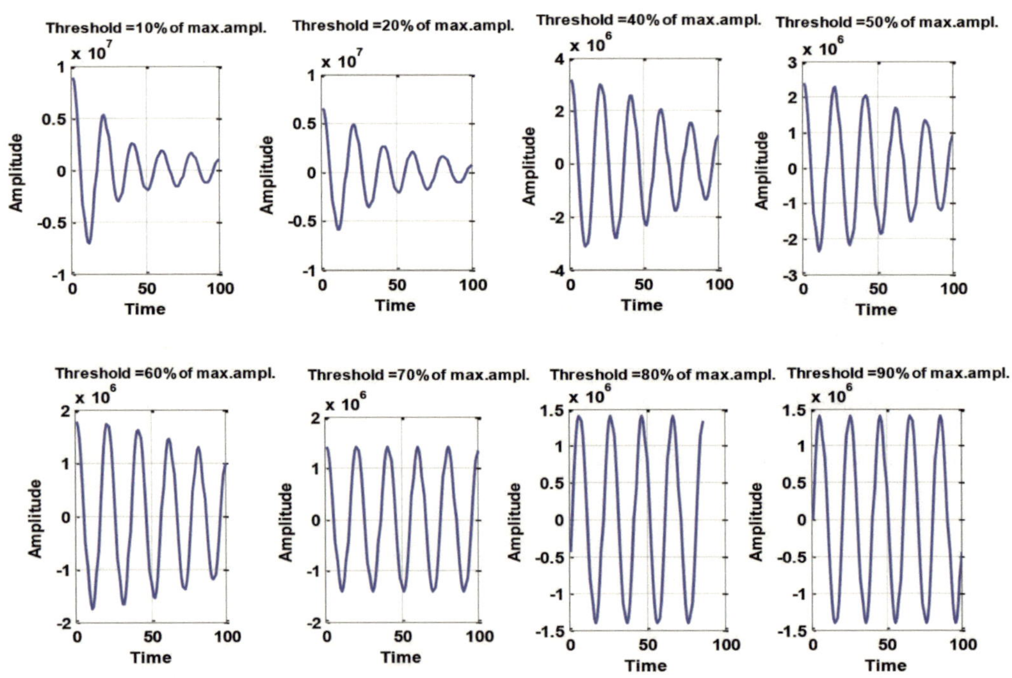

Figure 17. Denoised signal at different threshold.

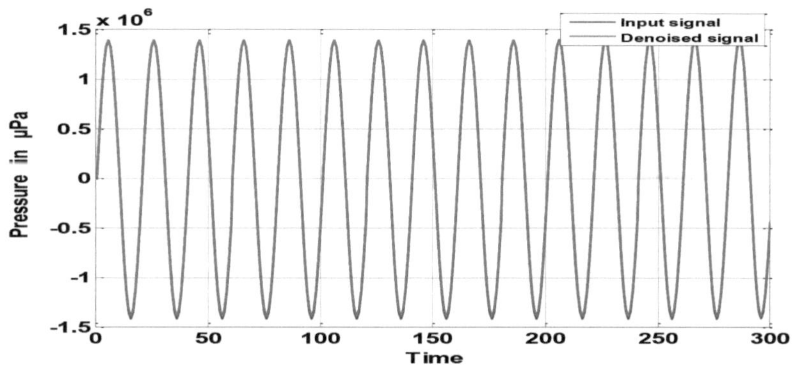

Figure 18. Input and denoised output signal comparison graph (EMD method).

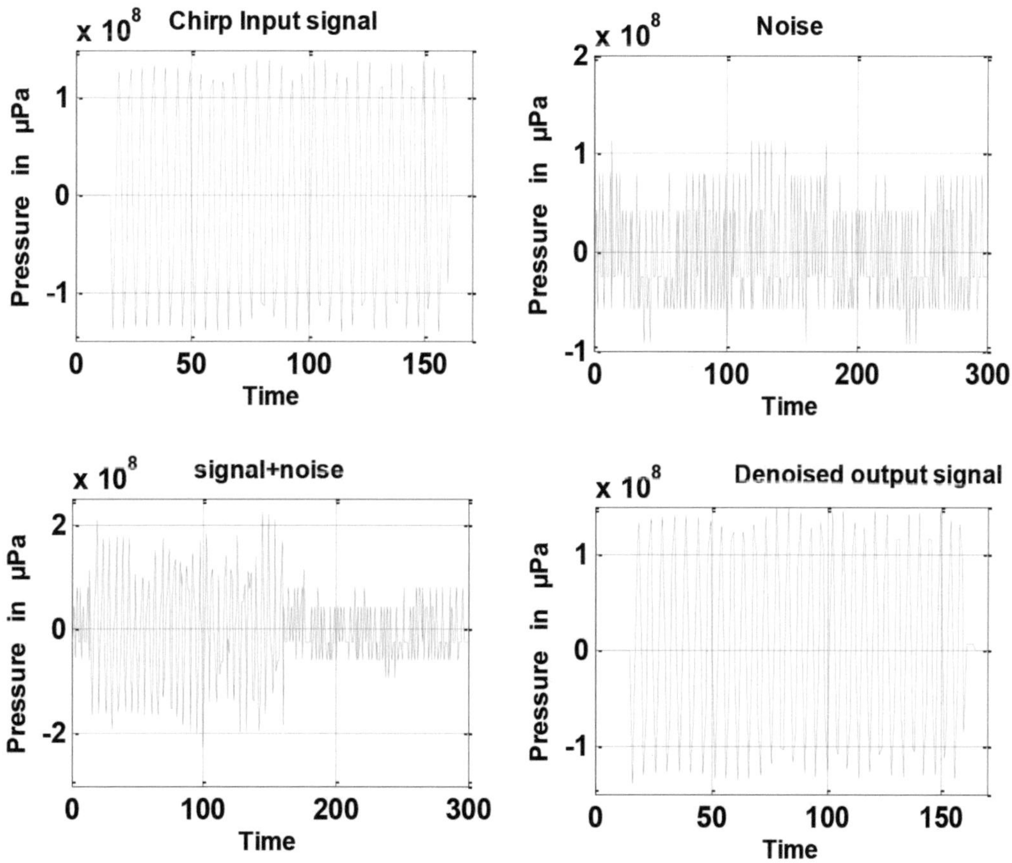

Figure 19. Output of proposed algorithm for chirp input signal.

In order to compare the performance of different algorithm, the RMSE value is calculated for different algorithm and various wind speed. It is presented in **Tables 2–5**. The RMSE value of the proposed algorithm is given in **Table 5** for different threshold values. From the table, it is

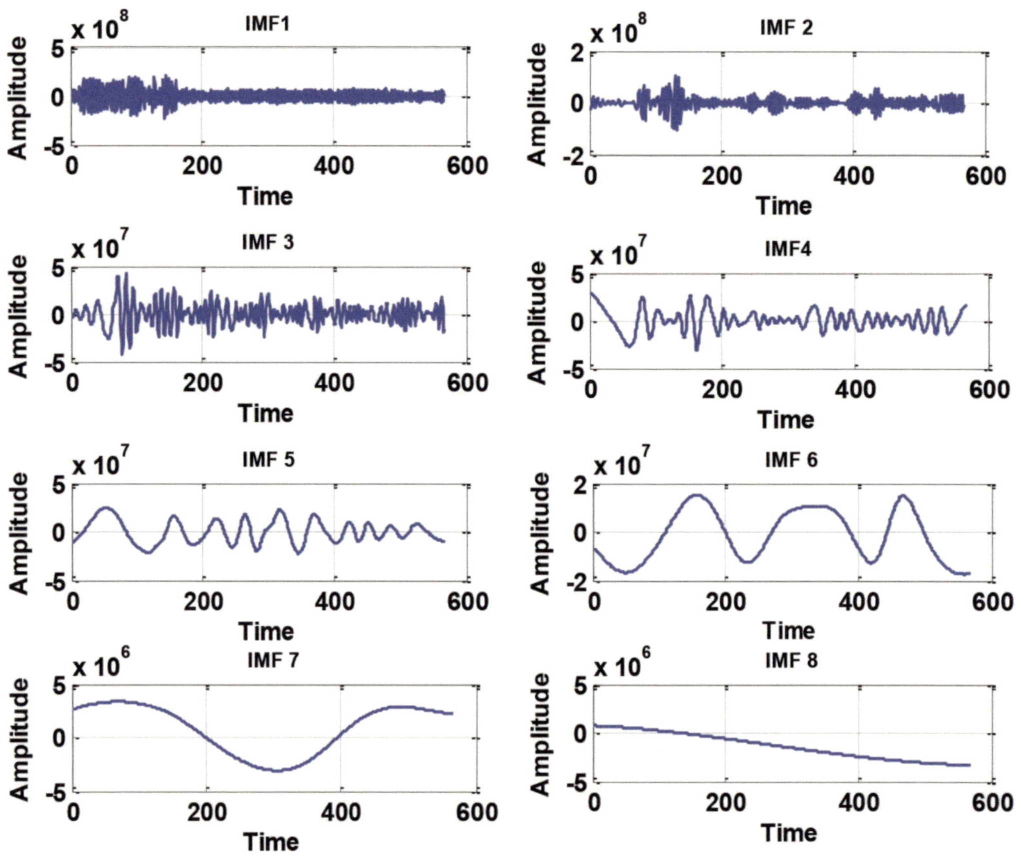

Figure 20. IMFs for chirp input signal.

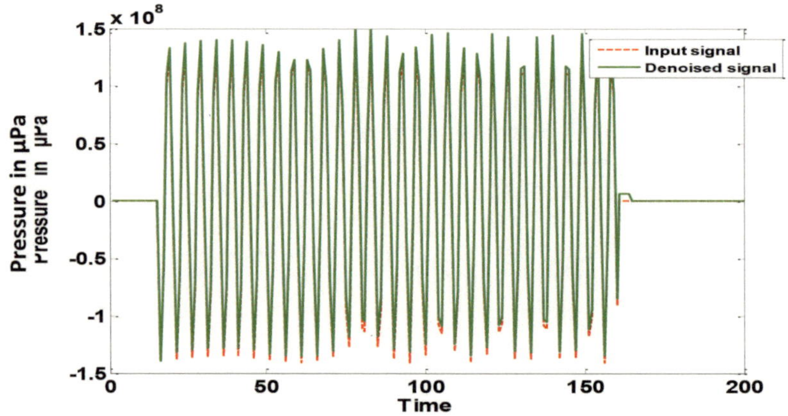

Figure 21. Input and denoised output signal comparison graph (for chirp input).

RMSE value at 2.61 m/s	RMSE value at 3.52	RMSE value at 5.06	RMSE value at 6.93 m/s
0.0029	0.00061	0.00088	0.0010

Table 2. RMSE in wavelet (signal amplitude = 0.01 mv).

RMSE value at 2.61 m/s	RMSE value at 3.52	RMSE value at 5.06	RMSE value at 6.93 m/s
0.0118	0.00141	0.0071	0.00821

Table 3. RMSE in time domain EMD (signal amplitude = 0.01 mv).

RMSE value at 2.61 m/s	RMSE value at 3.52	RMSE value at 5.06	RMSE value at 6.93 m/s
0.00060	0.00044	0.00063	0.00079

Table 4. RMSE in time domain using EEMD (signal amplitude = 0.01 mv).

Threshold value in %	RMSE value at 2.61 m/s	RMSE value at 3.52 m/s	RMSE value at 5.06 m/s	RMSE value at 6.93 m/s
20	0.002991	0.004051	0.002968	0.005838
40	0.000051	0.002512	0.002343	0.002191
50	0.000051	0.001906	0.001448	0.000780
60	0.000051	0.000919	0.001062	0.000780
70	0.000051	0.000045	0.001062	0.000064
80	0.000051	0.000045	0.000055	0.000064
90	0.0001675	0.000045	0.000055	0.000064

Table 5. RMSE in frequency domain EMD (signal amplitude = 0.005 mv).

clear that the RMSE value decreases as the threshold value increases, and there is no change in the RMSE value for the threshold above 80%.

7. Summary

The proposed frequency domain-based EMD algorithm outperforms all other existing algorithms. **Table 6** shows the RMSE, which is calculated for the wind noise signal of 5.06 m/s,

RMSE value for different denoising methods for wind noise signal of 5.06 m/s			
Wavelet	**EMD time domain algorithm**	**EEMD time domain algorithm**	**Proposed frequency domain based on EMD**
0.00088	0.0071	0.00063	0.000055

Table 6. RMSE value comparison for different denoising algorithm.

value for different denoising methods. The RMSE value in the proposed frequency domain algorithm is less compared to the other existing algorithms. The RMSE value is calculated for various wind speeds, and it consistently performs well in the proposed frequency domain approach. It is concluded that the proposed algorithm produces better results compared to the existing algorithm. "This algorithm has been developed and duly got tested for the wind noise and as such could be further extended to include the other constituents of the ambient noise, as well."

Acknowledgements

The author gratefully acknowledges the National Institute of Ocean Technology (NIOT), Chennai, for their support. He would also like to thank Dr. V. Rajendran, Professor, VELS University for his valuable guidance.

Author details

Vijaya Baskar Veeraiyan

Address all correspondence to: v_vijaybaskar@yahoo.co.in

Department of ETCE, Sathyabama University, Chennai, India

References

[1] Vijaya Baskar V, Rajendran V, Logashanmugam E. Study of different denoising methods for underwater acoustic signal. Journal of Marine Science and Technology. 2015;**23**(4):414–419

[2] Wenz GM. Acoustic ambient noise in the ocean: Spectra and sources. Journal of Acoustic Society of America. 1962;**34**:1936–1956

[3] Urick RJ. Ambient Noise in the Sea. USA: Peninsula Publishing; 1984

[4] Vijayabaskar V, Rajendran V. Wind dependence of ambient noise in shallow water of Arabian sea during pre-monsoon. In: Recent Advances in Space Technology Services and Climate Change 2010 (RSTS & CC-2010); pp. 372–375

[5] Etter PC. Underwater Acoustic Modeling: Principles Techniques and Applications. London: Elsevier Applied Science; 1991

[6] Chapman NR, Cornish JW. Wind dependence of deep ocean ambient noise at low frequencies. Journal of Acoustic Society of America. 1993;**93**:782–789

[7] Aggarwal R, Singh JK, Gupta VK, Rathore S, Tiwari M, and Anubhuti Khare A. Noise reduction of speech signal using wavelet transform with modified universal threshold. International Journal of Computer Applications. 2011;**20**:12–19

[8] Donoho DL. Denoising by soft-thresholding. IEEE Transactions on Information Theory. 1995;**41**:613–627

[9] Huang NE, Shen Z, Long SR, Wu MC, Shih EH, Zheng Q, Tung CC, Liu HH. The empirical mode decomposition method and the Hilbert spectrum for non-stationary time series analysis. Proceedings of the Royal Society. 1998;**454**:903–995

[10] Bajaj Varun, Ram Bilas Pachori BR. EEG signal classification using empirical mode decomposition and support vector machine. In: Proceedings of the International Conference on Soft Computing for Problem Solving (SocProS 2011), Springer India; 2011. pp. 623–635

[11] Boudraa AO, Cexus JC, Saidi Z. EMD-based signal noise reduction. International Journal of Information and Communications Engineering. 2005;**1**:96–99

[12] Flandrin P, Gonçalves P, Rilling G. Detrending and denoising with empirical mode decompositions. Proceedings of the 12th European Signal Processing Conference (EUSIPCO'04), IEEE. 2004;**2**:1581–1587

Chapter 5

Active and Passive Acoustic Methods for In-situ Monitoring of the Ocean Status

Sara Pensieri and Roberto Bozzano

Additional information is available at the end of the chapter

http://dx.doi.org/10.5772/intechopen.68998

Abstract

Recent European strategic plans for the successful monitoring of the status of the ocean push on the development of an integrated observing system able to further link existing instruments and techniques with the aim to complement each other and answer open issues. A more intensive use of acoustic devices could contribute to the knowledge of oceanographic processes exploiting the characteristic of sound to travel in the ocean for a wide area than in the atmosphere. In this context, the installation of passive acoustic instruments, able to listen to ambient noise on fixed or mobile platforms, could contribute to provide information on sound budget and to enhance the monitoring capacity of meteorological phenomena also in the open ocean. Instead, the deployment of active acoustic instruments can be of benefit for monitoring biological activities through the analysis of backscatter data as well as for monitoring ocean waves.

Keywords: active underwater acoustics, ocean passive underwater acoustic, underwater ambient noise, oceanography, in-situ monitoring

1. Introduction

A fully comprehensive picture of the ocean status can be obtained only by combining different methods and monitoring techniques exploiting the characteristics of each approach. The recent enhancement of remote sensing capabilities, in terms of a variety of measured parameters and accuracy of the corresponding estimates, has been contributing to an effective improvement of the skill of forecasting models and, generally speaking, to the whole domain of operational oceanography which also benefits in real-time measurements provided by equipped buoys, moorings, and mobile platforms (floats, AUV, glider, etc.) that represent the principal resources to acquire in-situ ocean measurements.

While methodologies for the monitoring of the physical properties of the ocean are fully assessed and guarantee a good spatial and temporal coverage, techniques to monitor biochemical processes as well as meteorological phenomena over the ocean are still open issues [1].

Underwater acoustic systems, both active and passive, could contribute to fill this gap by listening to the ocean noise or by transmitting pulses and interpreting the received echoes to improve the knowledge on biological activities or meteorological phenomena at sea such as rain and wind and the potential harmful impact of human activities on the ecosystem.

The feasibility of using underwater acoustics to propagate signals date back to 1918 but only during the Second World War there was a massive exploitation of devices to detect submarines through sound navigation and ranging system (sonar). The first civilian experiments to measure the sea bottom and to detect schools of fishes were carried out in the first half of the twentieth century [2]. Since then, the benefits of underwater acoustics were proportional to the technological developments in both hardware and software components, especially for oceanographic applications.

Several aspects have to be taken into account when planning an underwater acoustics measurement program. The most important factor is the type of noise being measured and, accordingly, its expected features in terms of amplitude, frequency, duration, and so on, which drive the choice of measurement equipment. Indeed, the sound in the ocean is characterized by speed of propagation, attenuation, and presence of obstacles along the path and by the way in which the sound is scattered, backscattered, and refracted by both the bottom and the surface.

Ocean stratification is the main responsible event for the generation of beams (convergent and/or divergent) and grey areas, depending on the change in the speed due to the depth. In turn, sea temperature profiles are influenced by diurnal cycle, season, and weather conditions. During winter months, the surface water that is colder and saltier tends to sink, and it is replaced by warmer and deeper water masses. This mixing could originate a layer of isotherm water characterized by a homogeneous sound velocity defined as "mixed layer." Below, the thermocline, that is the area in which temperature rapidly decreases with depth, dwells. The mixing of the water column implies an enlargement of the mixed layer and an erosion of the thermocline. The layer below the thermocline is characterized by a quite constant temperature and presents a minimum in the sound speed profile.

In the ocean, sound pressure levels (SPL) are retrieved using the sonar equation (Eq. (1)) as the difference between the transmitted power (SL) and the power loss (TL) through the path [3].

$$SPL\,[dB] = SL[dB] - TL[dB] = 10\log\left(\frac{P_0}{P_R}\right)^2 - \left[20\log\left(\frac{R}{R_0}\right) + \alpha\left(R - R_0\right)\right] \tag{1}$$

P_0 is the pressure of the transmitted signal at a known distance R_0, P_R is the reference pressure (generally equal to 1 μPa), and R is the distance of the listener from the source.

Equation (1) allows the quantification of SPL acquired by passive devices that simply listen to the numerous and heterogeneous ocean sounds like, among others, those produced by mammals, marine organisms, volcanoes, submarines, human activities, wind, waves, and rain.

Since active instruments are able to transmit a pulse and listen to its echo, it is necessary to consider also the intensity of the echo one meter from the target, that is, relative to the part of the sound that hits the target, the so-called target strength (TS). Thus, Eq. (1) is modified in Eq. (2)

$$SPL\ [dB] = SL[dB] - 2TL[dB] + TS\ [dB] = 10 \log \left(\frac{P_0}{P_R}\right)^2 - 2\left[20 \log \left(\frac{R}{R_0}\right) + \alpha\left(R - R_0\right)\right] + TS. \quad (2)$$

Among the others types of applications, active instruments are commonly used to detect schools of fish, mines, and currents.

In the design of an acoustic experiment also, the availability, or the construction, of adequate infrastructure for carrying out the measurements for a desired duration is a key requirement. The two most common approaches consist of using mobile and fixed platforms.

In vessel-based surveys, hydrophones (either individually or in arrays) are deployed from the ship, and the analysis and recording equipment remain on the vessel, which may be either anchored or drifting. This solution is relatively easy to implement, the deployments can be quick, a relatively large area may be covered, the risk of losing instrumentation is low, the configuration of hardware devices can be adjusted online, and data can be monitored in real time. Nonetheless, the main disadvantage consists of the pre-defined and limited (usually short) period of time during which the measurements can take place. Also, autonomous moving platforms such as gliders can be equipped with hydrophones to explore the soundscape of relatively large areas of the ocean.

When continuous time monitoring is of interest or when the objective is to observe episodic and non-predictable phenomena (i.e., biological and geological events), a Eulerian approach is preferable. This consists of the use of fixed observatories that can be based on sea bottom stations cabled to the shore [4] or on instruments deployed on oceanic sub-surface moorings [5] or surface buoys. Several large initiatives are currently operational all over the world: Ocean Observatories Initiative (OOI) in the USA, Neptune in Canada, European Seas Observatory NETwork (ESONET), and the neutrino telescope sites in Europe [6]. Cabled observatories allow data to be streamed directly to the shore base and checked in real time [7]. Unless the goal is to measure air-sea surface interactions through acoustics or characterize the acoustic signature of ships, bottom-mounted deployments offer the advantage of minimizing both the influence of surface wave action and the disturbance by surface vessels, reducing the risk to keeping the hydrophone away from the pressure-release water-air surface and the risk of damage to the equipment.

The characterization of the ambient sound all over the world oceans, through the variety of approaches mentioned above, has become more common as interest in the trends in anthropogenic sound in the ocean grows. The European Commission endorsed this issue considering the introduction of energy, including underwater noise, into the ocean as a pollutant [8] and requesting to monitor it with the same operative methodologies like other physical, biological, and chemical parameters.

2. Ocean waves statistics inferred by active acoustic devices

2.1. Ocean waves

Sea surface shows ripples of different dimensions and shapes, depending on the force of the wind speed, and on the basis of their characteristics, they can be subdivided in two categories: capillary and gravity waves. Ocean capillary waves are strictly connected to surface tension and show short wavelength, whereas ocean gravity waves are due to the force of the air-sea interface conditions and their wavelength can reach several meters, especially in open oceans during storms. The characteristics of the waves, induced by winds, are identifiable by wind speed intensity and distance and by the duration of the event [9, 10].

When energy loss due to the air-sea friction is negligible, waves can propagate until one of these events occur: wind forcing persists, waves are hindered by the presence of dams or consume their energy on the coastline. Dissipations of energy reduce inversely proportional to wavelength; thus, large wavelengths, generally faster, smooth slowly and propagate over long distance even where wind is absent.

Waves induced by wind force can be modeled by Eq. (3) as N_w sinusoids linearly interact each other, where each component is identified by an own amplitude (A_m), wavenumber (k_m), direction of propagation (θ_m), frequency (f_m), and initial phase(φ_m).

$$z(t) = \sum_{m=1}^{N_W} A_m \cos\left[k_m\left(x \cos\theta_m - y \sin\theta_m\right) - 2\pi f_m t + \varphi_m\right] \quad (3)$$

where x is the displacement along x axes, y the displacement along y axes, and φ is a random uniformly distributed variable between $-\pi$ and π. Sinusoidal waves with different frequencies propagate with the same speed related to bottom depth in shallow water, whereas they have a decreasing speed as the frequency rises in open ocean.

Wind blowing for an extended period of time over a long distance induces a rapid increase of both wave steepness and height. The upper limit of the height is reached when wave breaking generates a dissipation of energy able to balance the energy supplied by wind and, in this case, wave motion can be considered as fully developed. Each component of fully developed waves is a random ergodic process characterized by a variance equal to the mean quadratic value $A^2_m/2$. The variance of the whole wave field can be expressed as a summation of the N_w components in the Δf frequency band through Eq. (4) and it represents the monolateral power spectra.

$$S(f) = \frac{1}{2}\frac{\sum_{m=1}^{N_w} A_m^2}{\Delta f} \quad (4)$$

Through the 0th and 1st spectral moments of $S(f)$, it is possible to estimate the main characteristics of ocean waves. In fact, the mean pulse is the ratio between the 1st moment and the 0th moment, whereas the mean period is obtained dividing 2π by the mean pulse. The significant wave height can be defined as the average height of one-third of the highest measured waves ($H_{1/3}$) and can be expressed as 4.005 times the root square of the 0th moment [11].

2.2. Measuring ocean waves with acoustics

Several methodologies have been developed to estimate wave characteristics for open sea and coastal studies. In both environments, the difficulties to obtain measurements also with rough sea increased the use of data provided by satellite and, in the meantime, gave rise to a growing interest in the autonomous system capable to measure meaningful parameters in a continuous way and in all meteo-marine conditions.

In-situ technologies such as wave buoys [12], pressure and acoustic water level sensors [13], and upward-looking acoustic Doppler current profilers (ADCPs) [14] are generally employed to monitor and estimate ocean waves. Nonetheless, the use of a wave buoy is quite prohibitive in real open ocean environment with sea bottom deeper than 1000 m. In this case, the only possibility to collect wave estimate on long-term basis is to employ vertically oriented sonar installed on spar buoys [15, 16] that do not follow the surface but are designed to allow for negligible sensitivity to sea heave and height.

Acoustic wave meter systems are commonly based on a directional array of high frequency precision, and acoustic altimeters are installed in an upward-looking configuration. The echo-sounder transmits a short pulse, and the acoustic returns are amplified and subjected to compensation through a time-varying-gain circuit, which corrects for acoustic losses associated with beam spreading and attenuation in sea water. After digitization, the amplitudes of the echo are scanned to select a single target for each ping. The selection procedure chooses the target with the longest persistence from all targets having amplitudes above a user-specified threshold level.

Under the hypothesis of a constant sound speed, each altimeter emits a single beam toward the sea surface and measures the time between the emission and the received echo. Under stationary conditions of the sea state, wave height process can be considered as a stationary and ergodic stochastic process with zero mean. However, a truthful statistical description of sea waves is achieved only if the wave height process is supposed to be Gaussian [17].

In a real environment, not all samples satisfy the properties of the Gaussian distribution, and the measured echoes of the array of altimeters could be disturbed by reverberation of bubbles, dishomogeneity close to sea surface, and the presence of fishes lying between the altimeter and the sea surface. To overcome these issues, an ad-hoc processing algorithm has taken into account the correction for the motion of the platform hosting the acoustic array.

2.3. Wave meter system on spar buoys

An acoustic wave meter system was installed at a depth of 10 m on the spar buoy part of the W1M3A observatory moored in the open Ligurian Sea (Northwestern Mediterranean Sea) [18].

The array was constituted of three brackets, which were 2.5 m long, equally spaced at 120°, hosting three high-frequency (500 kHz) altimeters that emit a narrow conical beam (6.0° width at −3 dB) which results in a small area being insonified at the surface (about 1.04 m). A Transistor-Transistor Logic (TTL) signal triggered the emission of the pulse by each altimeter. In order to avoid interferences, the acquisition system, which controlled and collected the output signals from the altimeters, synchronized and slightly shifted in time with the three TTL signals so that each ping (and consequently each sample acquisition) was delayed with respect to the others of few milliseconds.

The slow motion of the spar buoy, especially in the presence of strong winds and currents, can influence the acoustic measurements, thus, the wave mater package was inclusive of vertical accelerometers and a couple of two axis orthogonal inclinometers installed along the horizontal axis to correct the acquired data for the buoy motion.

The acquisition system simultaneously collected the time series provided by the three echosounders at a frequency of 2 Hz and buoy motion data (inclination and acceleration). First, the time series of the three echosounders was quality controlled in order to identify spikes, outliers, and samples not satisfying the Gaussian condition. The detected samples were then interpolated by means of spline functions. In order to preserve the phase-shift information between the three time series in all the cases in which the reconstruction of part of the overall time series was not possible due to the elevated number of bad samples, all waveforms (provided by the altimeters, the inclinometer, and the accelerometer) were adjusted homogeneously.

The obtained time series was then filtered to compensate for platform motion [19], and standard statistical parameters (i.e., wave height, period) were computed on the basis of the spectral density features of the acoustic profiles [9, 20]. The wave meter system was designed to create an equilateral-triangular array (**Figure 1**), allowing the estimates of the prevalent direction of the wave by means of the theory of the direction of arrival [21] valid under the assumption of the incoming planar wave.

Let us consider to divide the three altimeters into pairs (i, j), (j, l), and (i, l): the time delay between the sensors of each pair when the planar wave passes through can be expressed in matrix form through Eq. (5):

$$\overline{\tau} = \overline{X} \cdot \overline{k}. \tag{5}$$

$\overline{\tau}, \overline{X}, \overline{k}$ represent the matrices of time delay, the displacement between the two devices of each pair, and the direction, respectively. If the triangularity condition between the three pairs (i, j), (j, l), and (i, l) of echosounder is satisfied $(x_{(i,j)} + x_{(j,l)} + x_{(i,l)} = 0)$, and it is known as the time delay vector, the propagation vector can be obtained by solving the last squares pseudo inverse Eq. (6).

$$\hat{k} = (X^T X)^{-1} X^T \hat{\tau}. \tag{6}$$

The assessment of the method was carried out comparing simultaneous wave estimates obtained by using the acoustic wave meter and acquired by a Datawell Waverider (DWR) directional buoy, a spherical one with 0.9 m of diameter, specifically designed to monitor wave characteristics. Wave data acquired by DWR buoys are basically displacement signals: one (the heave signal) for the non-directional wave rider and three (heave, north, and west displacement) for the directional wave rider. The mean, variance, skewness, and kurtosis of these signals are also computed. In the wave-statistical processing, zero-upcross waves are

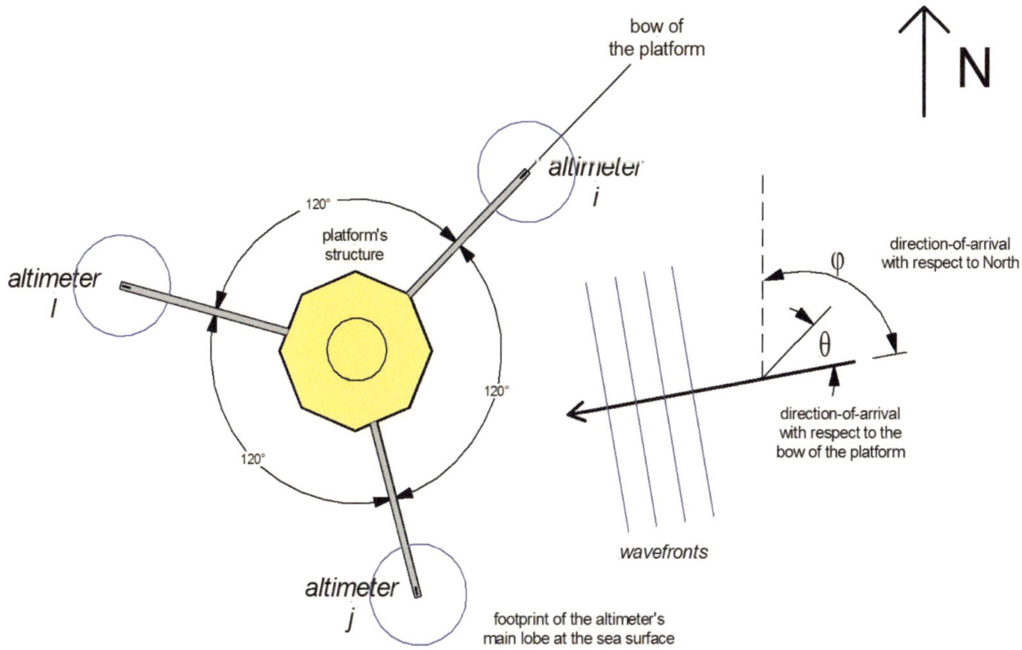

Figure 1. A sketch of the array of acoustic altimeters.

constructed from the heave signals, which are sorted by wave height and averaged in several fashions. This is the classic method of wave analysis that generates the significant wave height $H_{1/3}$. In the spectral analysis, the power spectral density is computed using Fourier methods. The wave direction as a function of wave frequency is calculated from the co-spectral and quadrature spectral densities of the three displacement signals. Using a maximum entropy method (MEM), the 3D spectrum, that is, the power spectral density as a function of both wave frequency and wave direction is computed.

The two buoys were moored at a nominal distance of 4 km for safety reasons, since the main buoy of the W1M3A observatory can span a circular area of 2 km by means of its slack mooring. These systems were continuously operational within the time of the validation which lasted 2 months from June to August, and all available estimates were used for the validation. During the period of the assessment, significant wave heights spanned from a minimum of 0.14 m to a maximum of 3.20 m and two storms occurred with the corresponding rough sea and strong wind speed. Thus, the acoustic wave meter was tested for several sea-state conditions. Although the majority of samples regard a smooth sea-state condition, a statistically significant number of samples refer to slight, and a moderate sea-state class was observed.

The validation of the acoustic wave meter system in terms of $H_{1/3}$ was based on the slope and the intercept of the linear regression line, considering the estimates are obtained using the acoustic method as the independent variable and the DWR observations as the dependent variable. Due to the constraints of the designed array, only sea wave with period greater than 3.3 s could be successfully measured without aliasing issues. The performed analysis shows

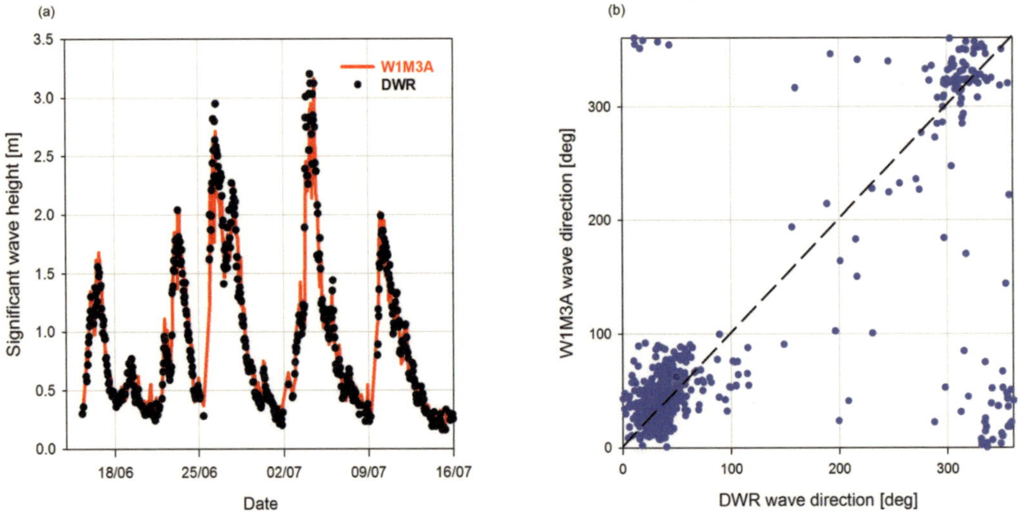

Figure 2. (a) Time series of significant wave height as estimated trough the acoustic wave meter system and measured by the Datawell Waverider buoy. (b) Scatter plot of the wave direction estimated by the acoustic wave meter versus the one measured by the Datawell Waverider buoy.

a very good agreement between the two time series (**Figure 2a**) with a correlation coefficient of 0.97 and a linear regression defined by a slope of 0.9017 and an intercept of 0.1052. The observed small variations can be considered as a feature of the natural sea state since the measurements were carried out in two different positions, about two nautical miles far. Wave direction estimates were compared considering the results of the direction of the arrival technique for the W1M3A data and the most powerful direction retrieved by the DWR directional spectrum. Results show an overall satisfactory agreement within the error of ± 15° that is consistent with the accuracy of the DWR estimates, except for few cases related to changes in wind direction and low wind intensity (**Figure 2b**).

Results demonstrate the feasibility to use an acoustic wave meter array as an affordable tool to measure waves on the long term and also in an open ocean where it is difficult to deploy discus buoys on deep sea bottom for an extended period of time. The system is still deployed on the W1M3A observatory and the collected data were used to indirectly assess the performance of the Dust Regional Atmospheric Model (DREAM) model to predict sea salt aerosol concentrations [22].

3. Migratory patterns of zooplankton detected by acoustic Doppler current profiler

3.1. Acoustic Doppler current profiler data

The first prototypes of acoustic Doppler current profilers were developed at the end of the 1980s with the aim of a continuous monitoring of ocean currents along the water column. Initially, these instruments transmitted a single narrowband pulse and through the auto-correlation technique and provided measurements of the first spectral order. Ten years later, the second generation of ADCP was put on market. It was characterized by a wider band and an enhanced data processing, exploiting the principle of Doppler effects. Nowadays, ADCP instruments transmit a pulse with known frequency and listen to the return echo that is backscattered from water drop, sediments, planktonic organisms, and all particles that are freely transported by ocean currents.

Part of the transmitted sound is backscattered in all directions, part is dissolved in the ocean, and another part comes back to the instrument. This signal is twice phase shifted because of the Doppler effect: when the scattering elements present in the ocean move away from the transducer, the sound is phase shifted of a quantity proportional to their relative speed (S) respect to the ADCP. If source and receiver are approaching or moving along a direction maintaining the same distance between them, no Doppler effect is present. For this reason, ADCP devices measure the parallel component to the acoustic beam, and the frequency of the signal turning back to the instrument (F_d) can be described through Eq. (7).

$$F_d = 2F_p\left(\frac{S}{c}\right)\cos\alpha \tag{7}$$

F_p is the frequency of the transmitted pulse, c is the sound speed in the ocean, and α is the angles between the beam and the water speed.

Mathematically, a phase displacement corresponds to dilation in the time domain. The sound produced by a single particle and also its backscatter echo remains unchanged until the particle doesn't move, but in case of a small displacement from the source, the echo will need more time to reach the transducer and thus the return signal will be phase shifted. ADCP devices measure the phase of the signal to obtain the time dilation exploiting the principle that the speed of the particles can be calculated if the interval of time between two pulses is known. The only ambiguity is represented by the fact that the phase is measured in the interval 0–360° and when the phase exceeds 360°, it starts again at 0°. The easy solution consists of transmitting a train of pulses with very short time delay for each pulse in order to avoid changes in the phase of the particles of more than 360°.

Generally, ADCP instruments are constituted of two couples of transducers to measure north, east, and vertical components of the ocean current, and the profile is obtained subdividing the water column in several segments called bins. The main outputs of the ADCP devices are current speed and direction, but several ancillary parameters, used to calculate current characteristics, are also available.

3.2. Diel vertical migration of zooplankton

The diel vertical migration (DVM) can be defined as the cyclic vertical displacement performed by most zooplankton species. Different DWM patterns have been observed, but the most common behavior is the swimming upward at sunset and downward at sunrise. Several environment causes such as light, temperature, food, and predation pressure, as well as endogenous origins like sex and age, influence DVM characteristics. Generally, the vertical distribution of the zooplankton is determined by net tows or pump samples that allow one to identify with the different species, but these samples are sparse in time and space and do not provide detailed information on the temporal variability, especially in the long-term period.

ADCP instruments are a powerful tool to overcome this issue, guaranteeing a quite continuous monitoring, also in extreme environments such as the Polar regions [23, 24] or highly productive basin, such as the Mediterranean Sea [25–27], at the expense of a specific taxonomic analysis.

Patterns of DVM can be detected through the analysis of the backscatter strength data (S_v) of the signal that, for the current profiler made by [28], can be expressed by Eq. (8).

$$\begin{gathered} S_v = 20\log R + 2\alpha R - A + 10\log\left(10^{\frac{K_c E}{10}} - 10^{\frac{K_c E_r}{10}}\right) \\ K_c = \frac{127.3}{T_x + 273} \\ R = \frac{B + \left(\frac{L+D+La}{2}\right) + [(n-1)D] + \frac{D}{4}}{\cos\vartheta} \end{gathered} \tag{8}$$

α is the absorption coefficient, T_x is the internal temperature of the device in °C, B is the distance in m beyond which the measure is valid, L is the length in m of the transmitted pulse, D is the dimension in m of the bin, L_a is the lag length of the pulses in m, n is the number of the cell in which the measure is taken, and ϑ is the inclination angle of the transducers. E is the raw echo signal as measured by the ADCP and E_r is the minimum acquired value during the deployment. R is defined as slant range and represents the spatial coefficient related to the inclination of the pulses with respect to the vertical of the instrument. The constant A is the best linear regression fit between Eqs. (8) and (9) proposed in Ref. [29] for all samples satisfying the condition of signal-to-noise ratio exceeding 10 ($K_c(E-E_r)<10$).

$$\begin{gathered} S_v = C + 10\log\left((T_x + 273.16)R^2\right) - L_{DBM} - P_{DBM} + 2\alpha R + K_c(E - E_r) \\ R = \frac{B + \left(\frac{L+D}{2}\right) + [(n-1)D] + \frac{D}{4}}{\cos\vartheta}\left(\frac{C'}{C_1}\right) \end{gathered} \tag{9}$$

C is a constant typical for each model of Teledyne RD Instruments (RDI) profiler and L_{BDM} and P_{BDM} are the logarithms of pulse length in m and of power transmission in Watt, respectively. C' is the sound velocity depending on the depth and C_1 is the sound speed used by the instrument to calculate the time between the pulse transmission and the received echo.

Equation (8) was applied to high-resolution acoustic ADCP backscatter data acquired during winter 2009–2010 in the Ligurian Sea. The used backscatter data were provided by an upward looking 300 kHz ADCP (by RDI) deployed at about 150 m depth on a deep sea bed of 1200 m from November 2009 to April 2010. The device was set to sample every 15 min with a bin length of 2 m in order to obtain high resolution data both in time and in space.

Backscatter strength values show a seasonal variability with low values in winter from 70 m depth down to 127 m and a gradual increase till 100 m in early spring in correspondence with the increment of the net primary productivity that, in turn, implies more availability of food in the water column and less need for the zooplankton population to reach the surface to feed themselves. The analysis of the S_v time series clearly evidences a circadian cycle with a quite perfect agreement between the decrease of backscatter intensity and the hour of sunrise (spanning from 5:08 to 06:38 UTC in the deployment period) and, vice versa, an increase of S_v in correspondence to the hours of sunset (ranging from 15:48 to 17:50 during the deployment). The analysis of corresponding vertical speed presents negative values around dusk and positive ones at dawn, showing a well-defined nocturnal DVM pattern with a rapid ascent of zooplankton from the bottom to the sea surface during dusk and a rapid descent from the surface to deep water at dawn (**Figure 3**).

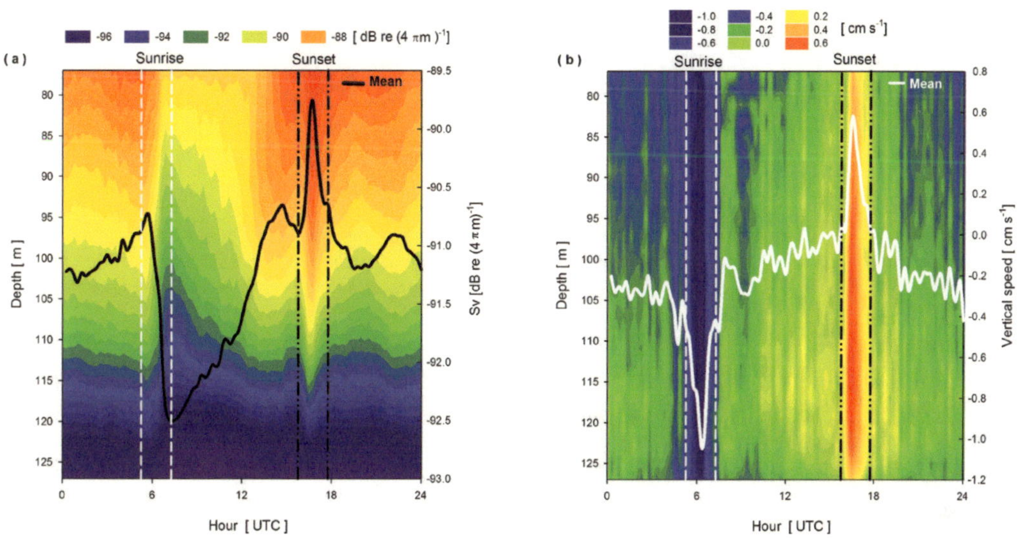

Figure 3. (a) Daily averaged backscatter strength with, superimposed, the mean values and (b) daily averaged vertical velocity with, superimposed, the mean values, the hours of sunrise and sunset.

Furthermore, in December and January, the DVM was influenced by moonlight: during full moon nights and clear skies (December 2 and 31, 2009), the backscatter strength decreased in the surface layers while greater values extended in the water column down to the maximum analyzed depth, making the values acquired at different depths quite homogenous along the water column (**Figures 4a** and **b**). This behavior, in contrast with the common nocturnal DVM of new moon periods (**Figures 4c** and **d**), is a characteristic of macrozooplankton/micronekton species and can be interpreted as a way to escape from visual predators [30]. Indeed, corresponding vertical velocities show more variability in the surface layers and a marked downward movement at midnight that is not present during the other moon phases (**Figures 4e** and **f**).

Obtained results demonstrate the feasibility to use non-calibrated ADCP data to infer zooplankton behavior with respect to daily seasonal and inter-annual variability as well as to astronomic phenomena. In fact, the observed intense DVM signal can be an indication of the presence of the *Clausocalanus* spp., *Fritillaria* spp., and, among the macrozooplanktons/

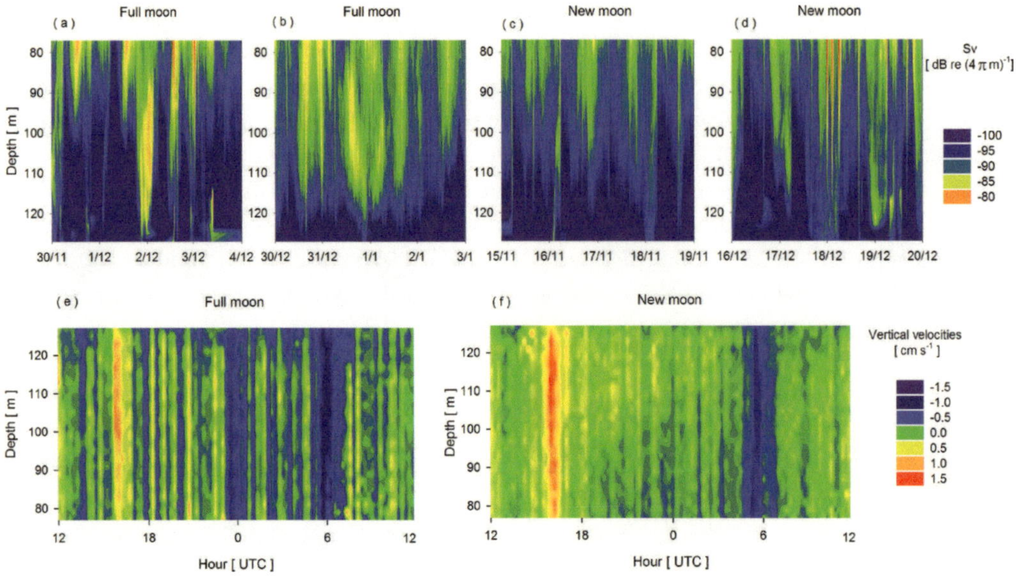

Figure 4. Temporal series of backscatter strength profiles during the full moon on (a) 2, December 2009 and (b) 31, December 2009 and (e) the corresponding average vertical velocities. Temporal series of backscatter strength profiles during the new moon on (c) 16, November 2009 and (d) 16, December 2009 and (f) the corresponding average of vertical velocities.

micronektons, of euphausiids (mainly *Meganyctiphanes norvegica*) that perform nocturnal migration and are abundant in the Ligurian Sea from December to March. Furthermore, the different patterns of DVM seen during full moon nights further support the hypothesis of the presence of euphausiids since, among the main migrators in the Ligurian basin, only euphausiids exhibit a sinking closely related to moonrise.

4. Ocean monitoring through passive acoustic measurements

4.1. Ocean environmental noise

Within the framework of the "Marine Strategy Directive to save Europe's seas and oceans" edited in June 2008, one of the main challenges of the Europe member state is to adopt mitigation actions and policy plans aiming at an effective protection of the overall marine environment by 2020. The increase of the maritime traffic and of anthropogenic activities at sea, such as the extensive use of sonar and oil drilling activities, has contributed to modify the natural ocean environmental noise so much that in some basins, it is the main cause of changes in the behavior of marine mammals.

Underwater environmental noise plays a fundamental role in biodiversity conservation, and the first studies date back to the Second World War when acoustic experiments established that environmental noise is the sum of several factors including ship traffic, breaking waves, wind, rain, mammals' vocalizations, and sound produced by marine organisms. In 1962,

Wenz [31] demonstrated that ships generate noise at low frequencies and proposed curves that describe the spectrum level at different frequencies for noise generated by ships and wind that were at the base, and still are, of forecasting systems. The National Research Council in Ref. [32] introduced a new definition of environmental noise as the "*noise associated with the background din emanating from a myriad of unidentified sources*." The most common sources can be distinguished by their acoustic signatures and can be subdivided into four major groups depending on their origin: physical, geological, biologic, and anthropogenic.

Wind is the major physical producer of noise over sea surface, and its spectral characteristics span a broadband frequency band, from less than 1 up to 50 kHz. The spectral curves show an increment for frequencies below 1 Hz, followed by a decrease as frequencies increase. As wind speed increases, the spectral curves maintain the same shape but show greater pressure levels. For wind speed > 10 ms^{-1}, the sound produced on the sea surface can be undistinguished by the sound due to the passage of a distant ship. Moreover, it is often associated with a high wave that is responsible for the generation of small bubbles that, in turn, produce sound and make the detection, and especially the quantification, quite difficult.

Also, precipitation contributes to the ocean noise in the frequency band from hundreds of Hz to more than 20 kHz, and the corresponding spectra show different characteristics depending on the type of precipitation. In the case of drizzle, a clear peak originated at the acoustic resonance of small drops splashing on the sea surface is observable around 15 kHz. This peak tends to disappear with the increase of the drops' dimension that produce sound at a frequency lower than 10 kHz and another peak at about 1–2 kHz in case of convective rain.

Tectonic processes, earthquakes, volcanic, and hydrothermal activities are the major geological sources contributing to the ocean environmental noise. Their spectra range from 1 to 100 Hz, show an initial burst, and the same noise persists for several minutes.

Biological sources are strictly related to marine organisms and mammals living in the ocean that produce signals spanning from 10 up to 200 kHz, depending on the species. In very productive basins, the biological sources are prevalent on the physical and geological components, whereas in high anthropological areas, the main responsible events of the noise are human activities.

Noise generated by ship passages is characterized by low frequencies (5–500 Hz) and propagates over long distances affecting wide areas. Each type of vessel (research vessels, leisure or fishing boats, tankers, commercial ferries, etc.) and also each single vessel are characterized by an own acoustic signature depending on cavitation phenomena, on the modulation of blade propeller, and on the on-board engines. Furthermore, noise produced by ships is variable and could be affected by environmental conditions especially for the interaction with the sea bottom.

Measurements of ocean environmental noise are related to the power of the propagating signal and to the characteristics of the acoustic path between source and receiver that can be modified by oceanographic dynamics, sound velocity propagation, and bathymetry. These components cause fluctuations in the pressure levels of the environmental noise depending on depth, time instants, and areas; thus, it is necessary to perform further experiments and to continuously monitor environmental noise to deepen the knowledge of its dynamic and the impact of the human activities.

4.2. Rainfall and wind speed measurements inferred by acoustic passive measurements

Passive acoustic data of ocean ambient noise consists of measurements of sound pressure as a result of the superimposition of sounds generated by several types of events (i.e., rainfall, ship passages, or mammals' vocalizations) to background noise, which is the natural noise in the absence of any sources, whose level is closely related to the intensity of blowing wind. Although background noise levels can be different from basin to basin, each source shows unique spectral characteristics that can be used to classify its type (physical, biological, anthropogenic) and, in some cases, also to obtain an estimate of atmospheric parameters over the sea surface (i.e., wind, rain). Indeed, multivariate analysis techniques can be applied to a combination of spectral levels, acquired at specific frequencies and least-square fit in different spectral bands to provide insights about the different sources forming the environmental noise.

Figure 5 shows the results of the multivariate analysis applied to acoustic data that was acquired in the open Ligurian Sea from March to November 2015 by means of a hydrophone installed on the W1M3A observatory. The output signal of the hydrophone was band-pass filtered and then digitalized at 16 bit with a sampling frequency of 100 kHz. Acoustic data

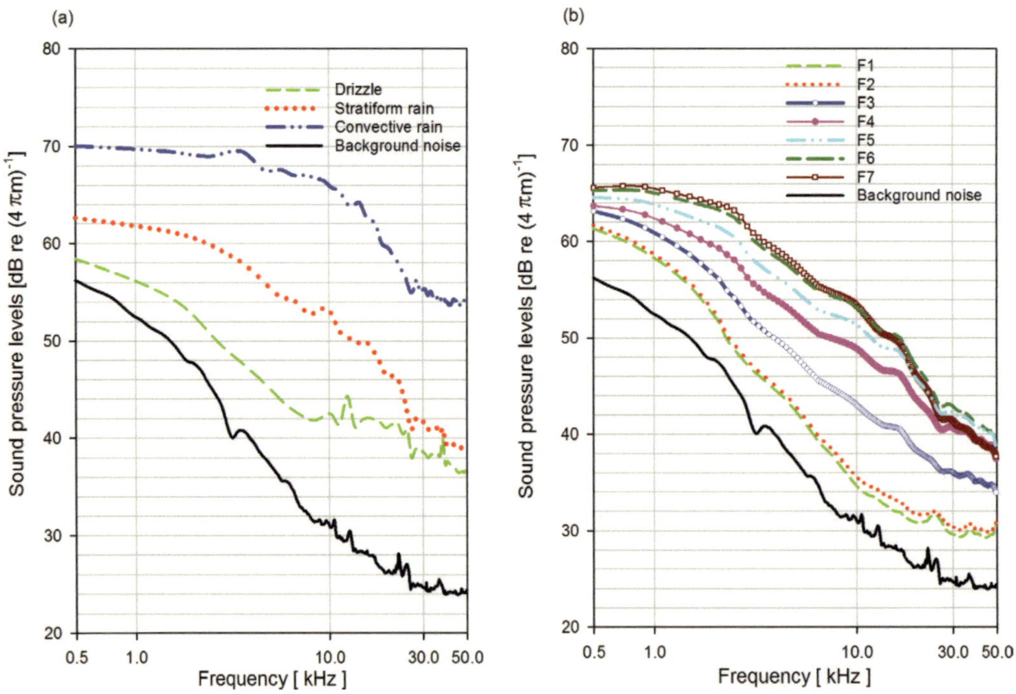

Figure 5. Average sound spectra for (a) different types of precipitation and (b) wind speed greater than 2 ms-1 subdivided in Beaufort classes compared to the average spectrum of the background noise.

were collected for few seconds, every tens of minutes, due to the large amount of data that such a sampling rate produces.

During the analyzed period, wind was continuously blowing over the sea surface and generated a sound that increased proportionally to the reinforcement of its speed, and, similarly, spectral levels tended to increase monotonically from 0.5 to about 25 kHz. Beyond this threshold, the sound produced by strong wind resulted comparable and even lower than the one generated by moderate breeze because of the contemporary arising of large waves and, in turn, the generation of small bubbles that absorbed the emitted sound. Wind spectra were very different to the one obtained during episodes of convective rain but could mask events of stratiform and light rain since the spectral levels from 20 up to 30 kHz were very similar for wind speed greater than 8 ms^{-1}. Furthermore, the resonance frequency of bubbles splashing on the sea surface is inversely proportional to their size and for this reason, large drops associated with heavy rain showed loud sound and, instead, small drops, typical of light rain events, presented a peak in the 10–15 kHz frequency band.

Several studies were carried out to quantify wind speed and rainfall amounts through the analysis of acoustic data. In Ref. [33], a logarithmic relation based on the sound pressure levels acquired at 8 kHz was proposed, and, recently, new parameterizations has been introduced for the Mediterranean region based on the results achieved during the Ionian Sea rainfall experiment and Ligurian sea acoustic experiment [34, 35].

The equation proposed in Ref. [35] was applied to the acquired acoustic data and compared to the in-situ wind speed observations provided by the W1M3A observatory for wind speed greater than 2 ms^{-1} (**Figure 6a**). Results show a good agreement between wind speed measurements provided by the anemometer and the estimates obtained using acoustic data, with a correlation of 87.5% and a root mean square error of 1.294 ms^{-1} taking into account that 2 ms^{-1} can be considered as the minimum wind speed that is acoustically detectable.

Rainfall rate and sound intensity are related by a logarithmic expression based on the sound pressure level at 5 kHz, whose coefficients can vary depending on the area of deployment [36]. Available acoustic data acquired in 2015 were processed following the algorithm proposed in Ref. [35] and compared to rainfall observations simultaneously acquired by a rain gauge installed on the W1M3A offshore observing system (**Figure 6b**). Results evidence the feasibility to use passive acoustic data to detect rainfall episodes, especially in case of intense events and the capability of quantity rainfall amounts with good accuracy, independently from rain types and the presence of wind speed.

4.3. Marine mammals monitoring through passive acoustic observations

Passive acoustic observations provide powerful support to complement visual surveys for the monitoring of marine mammals due to the fact that acoustic waves propagate for long distance. Visual observations are weather dependent, not available in remote or inaccessible areas, often limited in their use due to the short times animals may spend at the surface, and are sparse in time, whereas passive acoustic devices can be successfully employed for an extended period of time and can monitor a wide area providing information about both the presence and the species of the animals.

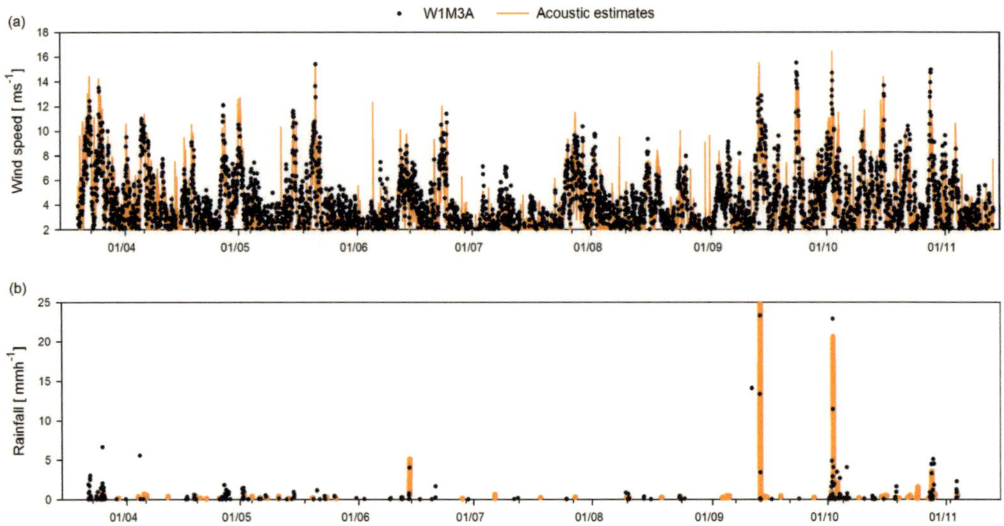

Figure 6. Time series of (a) wind speed and (b) rainfall as measured by the anemometer and the rain-gauge installed on the W1M3A marine observatory, and the estimates obtained from acoustic samples.

Vocalizations are the principal sounds generated by marine mammals that use them to communicate, to echolocate, and also for predatory or mating purposes [37]. Every species, and even each individual, can be recognized by its acoustic signatures, and for this reason, time-frequency analysis of time series of passive acoustic data is useful for the marine mammals' monitoring. This is particularly true for basins where human activities are scarce and the environmental noise is dominated by mammals' vocalizations like Thetys Bay in Antarctica or in a very productive area such as the "Cetacean Sanctuary" in the Ligurian Sea.

During the 29thItalian Antarctic expedition, a hydrophone was installed under the sea ice in Thetys Bay to study sound propagation. The basin is a natural habitat of different pinnipeds species (i.e., crabeater seal (*Lobodon carcinophaga*), leopard seal (*Hydrurga leptonyx*), Ross seal (*Ommatophoca rossii*), and Weddel seal (*Leptonychotes weddellii*)), and the vocalizations of some of these mammals were the preponderant sound in the collected measurements (**Figure 7**).

The experiment took place in November, during the Weddel seals mating period [38], and this explains the reason why the prevalent types of detected calls from Weddel seals are trills and whistles, both ascending or descending as defined in Ref. [39]. Trill calls show a descending pattern, are emitted once, last for 15 s and cover a wide frequency range from 6 kHz down to few hundreds of Hz. Whistles ascending, although being single pulses, last few seconds maximum, and their patterns are characterized by a sharp increase from about 4 up to 5 kHz, followed by a smooth rise up to 6 kHz maximum. Whistles descending are a series of pulses initially emitted at about 1-s intervals, progressively reducing the interval and dropping from 10 to 2 kHz. The typical vocalization produced by crabeater seal is known as moan; its spectrum has power content lower than trills and whistles in the whole range of frequencies and the signal spans from 700 up to 6 kHz.

http://dx.doi.org/10.5772/intechopen.68998

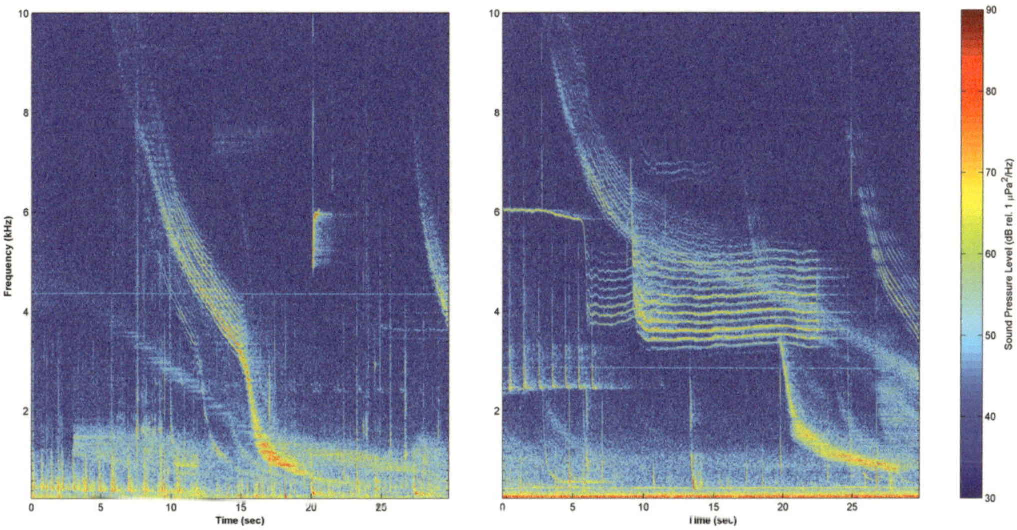

Figure 7. Spectrograms of vocalizations by Weddel and crabeater seals acquired during the 29th Italian Antarctic expedition.

A similar experiment took place in the Ligurian Sea during 2015, allowing one to identify the presence of sperm whales (*Physeter macrocephalus*), striped dolphins (*Stenella coeruleoalba*), and Cuvier's baked whales (*Ziphius cavirostris)* through the analysis of their spectrograms (**Figure 8**).

Sperm whales are the most common mysticete species in the Ligurian-Corsican-Provençal basin due to the high productivity that characterized the area supported by the permanent frontal structure of rich large biomass of krill, especially of *Meganyctiphanes norvegica* that is the favorite prey of the sperm whales. The vocalizations of sperm whale are constituted of sequences of clicks, which are brief impulsive sounds, variable in length that can reach 35 kHz in frequency. The pattern is depending on the area, the sex, the age of the animal, and also on the meaning: train of pulses with a repetition rate of two to three clicks per second are emitted during the diving to make recognition of the environment or for hunting, whereas high rate clicks referable to creaks are commonly used for echolocation.

Odontocetes calls are much different from mysticete's vocalization, presenting a wide variety of patterns of whistles ranging from few Hz up to more than 20 kHz and clicks used for echolocation that can extend between 50 and 150 kHz. Using the spectrograms, it is possible to distinguish the different species of the odontocetes and, in some cases, the sound emitted by the same individual.

The availability of passive acoustic recordings covering a long period of time could really improve the knowledge of mammals' vocalizations in their natural environment, especially in winter months where it is difficult to carry out visual surveys due to potential bad weather conditions.

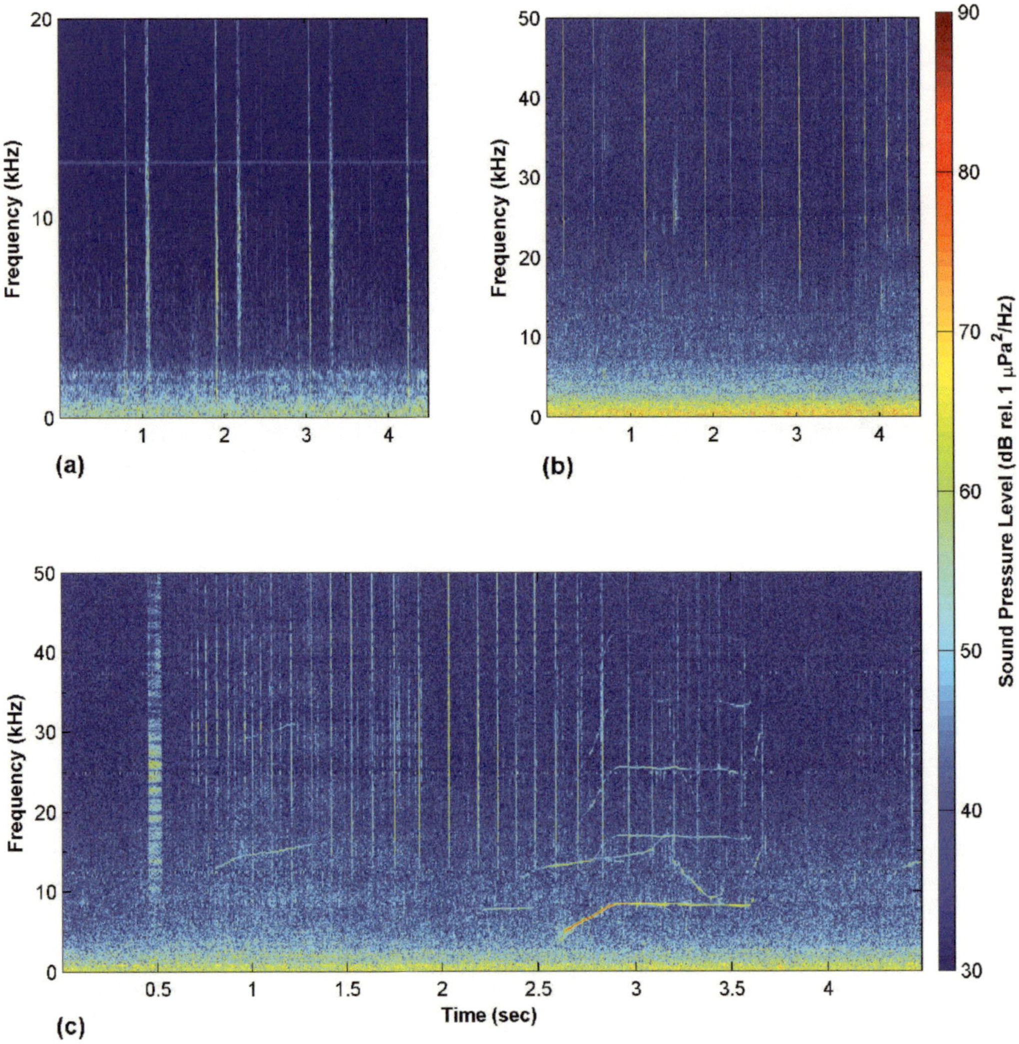

Figure 8. Spectrograms of vocalization by (a) sperm whales, (b) Cuvier's baked whales, and (c) striped dolphins acquired in the Ligurian Sea.

5. Summary

The combination of active and passive underwater acoustic methods could significantly contribute to the monitoring of the oceanic environment and to a better characterization of the ocean status. Analysis of acoustic observations in the time domain allows the detection of seasonal trends or inter-annual variability helpful for the identification of climate change's causes and/or impacts, as well as for the definition of mitigation actions and strategic plans devoted to the protection of the marine environment. Otherwise, analysis of acoustic data in the frequency domain makes possible to distinguish geophysical phenomena, such as wind and rain, and biological sources, such as vocalizations of marine mammals and anthropogenic noise by means of their own acoustic signatures. Specifically, the application of Fast Fourier Transform (FFT), wavelet, and autocorrelation techniques could provide insights about wave fields and give evidence of the presence of several marine mammals or different patterns referable to migratory processes, typical of zooplankton and micronekton species.

Indeed, in-situ acoustic measurements provided by a directional array of upward looking echosounders, installed on a spar buoy, have been used to obtain estimates of wave height, period, and direction in the open Ligurian Sea. Results show the feasibility to use acoustics to obtain reliable observations of wave field using a fixed platform not specifically designed to follow the slope of the waves. Measurements provided by active devices have been also successfully employed to monitor the behavior of zooplankton in relation to daily cycle and moon illumination for a long period of time that cannot be obtained using sporadic cruises or net samples sparse in time.

Experiments based on the installation of hydrophones carried out in different basins demonstrated the potentiality of passive acoustic data used to identify a variety of processes. Known as the mean noise level of the basin in which the hydrophones are deployed, it was possible to apply algorithms to automatically quantify rain and wind by means of the noise produced on the sea surface. Furthermore, the application of time-frequency analysis allowed the creation of spectrograms from which the types of mammals living in a different basin were easily detected.

Acknowledgements

The research leading to these results has received part of the funding from the European Community's Seventh Framework Programme FP7/2007–2013 under the grant agreement n° 312463 (FixO3) and the Flagship Project RITMARE funded by the Italian Ministry of Education, University, and Research.

Author details

Sara Pensieri* and Roberto Bozzano

*Address all correspondence to: sara.pensieri@ge.issia.cnr.it

Institute of Studies on Intelligent Systems, National Research Council of Italy, Genova, Italy

References

[1] Zampoukas N, Piha H, Bigagli E, Hoepffner N, Hanke G, Cardoso AC. Monitoring for the Marine Strategy Framework Directive: Requirements and Options. European Commission—Joint Research Centre, Institute for Environment and Sustainability, Luxembourg; 2012. DOI: 10.2788/77640

[2] Sund O. Echo sounding in Fisheries Research. Nature. 1956;**135** p.593. doi:10.1038/135953a0

[3] Urick RJ. Principles of Underwater Sound. 3rd ed. New York: McGraw-Hill, Inc.; 1983. p. 423

[4] Wall Bell CC, Rountree RA, Juanes F. Mapping the acoustic soundscape off Vancouver Island using the NEPTUNE Canada Ocean Observatory. Advances in Experimental Medicine and Biology. 2016;**875** p. 1205-1211, doi: 10.1007/978-1-4939-2981-8_151

[5] Erbe C, Verma A, McCauley R, Gavrilov A, Parnum I. The marine soundscape of the Perth Canyon. Progress in Oceanography Part A. 2015;**137**:38-51. DOI: 10.1016/j.pocean.2015.05.015

[6] Adam O, Glotin H. Passive acoustic storey of the Antares neutrino detector for real-time cetaceans detection, localization and behavior studies. In: New Trends for Environmental Monitoring Using Passive Systems; Institute of Electrical and Electronics Engineers (IEEE), Hyeres, French Riviera. 2008. pp. 1-6. DOI: 10.1109/ PASSIVE.2008.4786981

[7] André M, van der Schaar M, Zaugg S, Houégnigan L, Sánchez AM, Castell JV. Listening to the deep: Live monitoring of ocean noise and cetacean acoustic signals. Marine Pollution Bulletin. 2011;**63**(1-4):18-26. DOI: 10.1016/j.marpolbul.2011.04.038

[8] Tasker ML, Amundin M, André M, Hawkins A, Lang W, Merck T, Scholik-Schlomer A, Teilmann J, Thomsen F, Werner S, Zakharia M. Marine Strategy Framework Directive—Task Group 11 Underwater noise and other forms of energy. EUR—Scientific and Technical Research Series ed. Joint Research Luxembourg, Centre; 2010. p. 55. DOI: 10.2788/87079

[9] Earle MD, Bishop JM. A Practical Guide to Ocean Wave Measurement and Analysis. Marion: Endeco Inc.; 1984

[10] Massel SR. Ocean Surface Waves; their Physics and Prediction. London: World Scientific Publication; 1996

[11] Holthuijsen LH. Waves in Oceanic and Coastal Waters. Cambridge: Cambridge University Press; 2007. p. 387

[12] Varstow SF, Ueland G, Krogstad HE, Fossum BA. The Wavescan second generation directional wave buoy. IEEE Journal of Oceanic Engineering. 1991;**16**(3):254-266

[13] Weiergang J. Wave measurement by single beam acoustic profiling. In: Proceedings of the ECUA 1992; 14-18 September: Elsevier Applied Science; Luxembourg. 1992. pp. 157-160

[14] Terray EA, Grodon RL, Brumley B. Measuring wave height and direction using upward-looking ADCPs. In: Proceedings of the IEEE OCEANS '97; 6-9 October; Halifax. 1997

[15] Graber HC, Terray EA, Donelan MA, Drennan WM, Van Leer JC, Peters DB. ASIS—a new Air–Sea Interaction Spar buoy: Design and performance at sea. Journal of Atmospheric and Oceanic Technology. 2000;**17**:708-720

[16] Zedel L. Deep ocean wave measurements using a vertically oriented sonar. Journal of Atmospheric and Oceanic Technology. 1994;**11**:182-191

[17] Rychlik I, Johannesson P, Leadbetter MR. Modelling and statistical analysis of ocean-wave data using transformed Gaussian processes. Marine Structures. 1997;**10**:13-47

[18] Canepa E, Pensieri S, Bozzano R, Faimali M, Traverso P, Cavaleri L. The ODAS Italia 1 buoy: More than forty years of activity in the Ligurian Sea. Progress in Oceanography. 2015;**135**:48-63. DOI: 10.1016/j.pocean.2015.04.005

[19] Anctil F, Donelan MA, Drennan WM, Graber HC. Eddy correlation measurements of air–sea fluxes from a discus buoy. Journal of Atmospheric and Oceanic Technology. 1994;**11**:1144-1150

[20] Rychlik I. A note on significant wave height. Oceanic Engineering. 1996;**23**(6):447-454

[21] Pirinen TW, Yli-Hietanen J. Time delay based failure-robust direction of arrival estimation. In: 2004 Sensor Array and Multichannel Signal Processing Workshop Proceedings; Institute of Electrical and Electronics Engineers (IEEE), Barcelona, 2004. pp. 618-622. DOI: 10.1109/SAM.2004.1503023

[22] Kishcha P, Starobinets B, Bozzano R, Pensieri S, Canepa E, Nickovic S, di Sarra A, Udisti R, Becagli S, Alpert P. Sea-salt aerosol forecasts compared with wave height and sea-salt measurements in the open sea. In: Down GS, Trini Castelli S, editors. Air Pollution Modeling and its Applications XXI. NATO Science for Peace and Security Series-C: Environmental Security. Dordrecht, Springer; 2012. pp. 299-303. DOI: 10.1007/978-94-007-1359-8_51

[23] Cisewski B, Strass VH, Rhein M, Krägefsky S. Seasonal variation of diel vertical migration of zooplankton from ADCP backscatter time series data in the Lazarev Sea, Antarctica. Deep-Sea Research Part I: Oceanographic Research Papers. 2010;**57**(1):78-94. DOI: 10.1016/j.dsr.2009.10.005

[24] Picco P, Schiano ME, Pensieri S, Bozzano R. Time-frequency analysis of migrating zooplankton in the Terra Nova Bay polynya (Ross Sea, Antarctica). Journal of Marine Systems. 2017;**166**:172-183. DOI: 10.1016/j.jmarsys.2016.07.010

[25] van Haren H. Internal wave–zooplankton interactions in the Alboran Sea (W-Mediterranean). Journal of Plankton Research. 2014;**36**(4):1124-1134. DOI: 10.1093/plankt/fbu031

[26] Bozzano R, Fanelli E, Pensieri S, Picco P, Schiano ME. Temporal variations of zooplankton biomass in the Ligurian Sea inferred from long time series of ADCP data. Ocean Science. 2014;**10**:93-105. DOI: 10.5194/os-10-93-2014

[27] Pinot J, Jansá J. Time variability of acoustic backscatter from zooplankton in the Ibiza Channel (Western Mediterranean). Deep Sea Research Part I. 2001;**48**:1651-1670

[28] Gostiaux L, Van Haren H. Extracting meaningful information from uncalibrated backscattered echo intensity data. Journal of Atmospheric and Oceanic Technology. 2010;**27**(5):943-949

[29] Deines L. Backscatter estimation using broadband acoustic Doppler current profilers. In: Oceans 99 MTS/IEEE Conference Proceedings; 13-16 September 1999; Institute of Electrical and Electronics Engineers (IEEE), San Diego, CA, USA

[30] Tarling GA, Buchholz F, Matthews JBL. The effect of a lunar eclipse on the vertical migration behaviour of Meganyctiphanes norvegica (Crustacea: Euphausiacea) in the Ligurian Sea. Journal of Plankton Research. 1999;**21**(8):1475-1488

[31] Wenz GM. Acoustic ambient noise in the ocean: Spectra and sources. Journal of the Acoustical Society of America. 1962;**34**:1936-1956

[32] National Research Council, editor. Ocean Noise and Marine Mammals. National Academies Press; Washington, D.C. 2003

[33] Vagle S, Large WG, Farmer DM. An evaluation of the WOTAN technique of inferring oceanic winds from underwater ambient sound. Journal of Atmospheric and Oceanic Technology. 1990;**7**:576-595. DOI: 10.1175/1520-0426(1990)007<0576:aeotwt>2.0.co;2

[34] Anagnostou MN, Nystuen JA, Anagnostou EN, Nikolopoulos EI, Amitai E. Evaluation of underwater rainfall measurements during the Ionian Sea rainfall experiment. IEEE Transactions on Geoscience and Remote Sensing. 2008;**46**(10):2936-2946. DOI: 10.1109/tgrs.2008.2000756

[35] Pensieri S, Bozzano R, Nystuen JA, Anagnostou EN, Anagnostou MN, Bechini R. Underwater acoustic measurements to estimate wind and rainfall in the Mediterranean Sea. Advances in Meteorology. 2015;**2015**:612512. DOI: 10.1155/2015/612512

[36] Nystuen JA. An explanation of the sound generated by light rain in the presence of wind. In: Kerman BR, editor. Natural Physical Sources of Underwater Sound. Kluwer Academic Publishers; Dordrecht, 1993. pp. 659-668

[37] Verfuss UK, Miller LA, Schnitzler HU. Spatial orientation in echolocating harbour porpoises (*Phocoena phocoena*). Journal of Experimental Biology. 2005;**208**(17):3385-3394

[38] Doiron EE, Rouget PA, Terhune JM. Proportional underwater call type usage by Weddell seals (Leptonychotes weddellii) in breeding and nonbreeding situations. Canadian Journal of Zoology. 2012;**90**(2):237-247

[39] Thomas JA, Kuechle VB. Quantitative analysis of Weddell seal (Leptonychotes weddelli) underwater vocalizations at McMurdo Sound, Antarctica. Journal of the Acoustical Society of America. 1982;**72**:1730-1738

Chapter 6

Development of Quantitative Single Beam Echosounder for Measuring Fish Backscattering

Henry M. Manik, Dony Apdillah,
Angga Dwinovantyo and Steven Solikin

Additional information is available at the end of the chapter

http://dx.doi.org/10.5772/intechopen.69156

Abstract

Target strength (TS) of marine fish is a key factor for target identification and stock quantification. Validation of measurement and model comparisons in fisheries acoustics is difficult, due to the uncertainty in ground truth obtained in the ocean. To overcome this problem is to utilize laboratory measurements, where fish parameter is more well controlled. In this research, the dorsal-aspect TS of fish was measured as a function of the incidence angle in a water tank using a quantitative echo sounder. The measurement was compared with the theoretical prediction using the distorted-wave born approximation (DWBA) model. TS of fish was proportional to body length and the directivity of TS was strongly dependent on its orientation. Computational DWBA modeling, experimental details, and data/model comparison were presented.

Keywords: target strength, high resolution, sonar equation

1. Introduction

Underwater acoustics technologies are frequently used to measure the abundance and biomass of fish [1]. The quantitative relationship between the size of a fish and its target strength (TS) and the intensity of the echo returned from the fish are important [2]. The swim bladder of fish is responsible for most of the reflected sounds [3]. TS of fish was determined also by size and shape of swim bladder [4, 5]. The acoustic target strength of a fish is required to enable the performance of present and future sonar equipment to be determinates for fish targets. Target strength is a logarithmic measure of the energy scattered by an object back toward the source and is a function of the size, shape, orientation, and material properties of the target [6].

A physical-based model of the acoustic scattering from the targets is required to convert acoustic backscatter measurements into units of fish density and biomass [7]. The physics-based scattering model requires input parameters describing the acoustic frequency of echo sounder system and the target (shape, length, orientation relative to the acoustic wave, and material properties) [8]. The properties of fish for acoustic modeling are ratio of fish density and seawater density (g) and ratio of the speed of sound in fish and the sound speed of seawater (h) [9, 10]. One purpose of this study was to examine the influence of material properties, specifically g and h, on model predictions of fish target strength (TS).

2. Material and methods

2.1. Measurement of fish target strength

Acoustic data were collected in the water tank of the Ocean Acoustics Laboratory Department of Marine Science and Technology Bogor Agricultural University. The echo sounder used in the studies was 200-kHz single-beam SIMRAD EK15. For the numerical model of distorted-wave born approximation (DWBA) purpose, we combine this instrument with 50 kHz. The specification of single-beam echo sounder was shown in **Table 1**. The echo sounder was calibrated with standard

Frequency [kHz]	200
Pulse duration [μs]	80
Ping rate [Hz]	40
Ping interval [ms]	500
Beam width [degrees]	26
Output power [W]	45
Bandwidth [Hz]	3088

Table 1. Specification of single-beam echo sounder Simrad EK15.

copper spheres as recommended by the manufacturer. The program designed was used to calibrate the single-beam units. Single-beam data were analyzed using Sonar 5 software (developed by Helge Balk and T. Lindem, Institute of Physics, the University of Oslo, Norway) and Matlab. This program used the algorithm to derive fish target-strength distributions from the measured distribution of peak voltage response from single-fish echoes (40 log R TVG function) [11]. Single-fish echoes are defined as echoes with less than twice the pulse length [11]. Due to the echo sounder-hardware noise and software limitation, we used −55 dB as the smallest target-strength group for the single-beam sonar. The method provides information for species identification, makes it possible to measure the fish length of individual fish, and provides information on fish behavior. Flow of research was shown in **Figure 1**. Beam pattern of transducer $B(\theta)$ is plotted on a decibel scale where the sound pressure as a function of spherical angle is

$$B(\theta) = 20 \log \left[\frac{2 J_1(\pi \frac{D}{\lambda}) \sin \theta}{\pi \frac{D}{\lambda} \sin \theta} \right] \tag{1}$$

θ is the angle of sound pressure from an axis perpendicular to the transducer center, D is transducer diameter, λ is wavelength of the sound, and $J_1$is first order Bessel function.

2.2. Physic-based scattering model

The theoretical scattering model used was distorted wave born approximation (DWBA). The DWBA model was originally used for weak scatterers such as zooplankton and micronecton. However, it has also been applied to fish. The DWBA model is valid for all acoustic frequencies, can be evaluated for all angles of orientation [12, 13], and can be applied to arbitrary shapes. DWBA model is valid when the incident acoustic wave is higher than the scattered value. Formulation of this model involved the incident acoustic wave number inside the integral. The amplitude of fish backscattering is given by

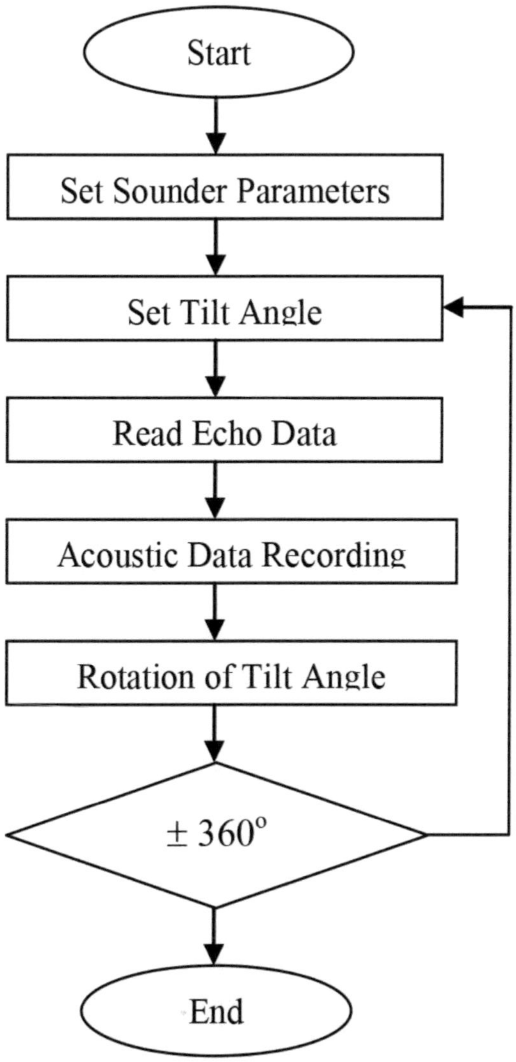

Figure 1. Flowchart of data acquisition system.

$$f_{bs} = \frac{k_1^2}{4\pi} \iiint_V (\gamma_\kappa - \gamma_\rho) e^{1\, i\, k_2 . r_0} dV \tag{2}$$

The terms γ_k and γ_ρ are compressibility k and ρ, and subscript v is parameter of the scattering volume.

$$\begin{aligned} \gamma_\kappa &\equiv \frac{\kappa_2 - \kappa_1}{\kappa_1} = \frac{1 - g\, h^2}{g\, h^2} \\ \gamma_\rho &\equiv \frac{\rho_2 - \rho_1}{\rho_2} = \frac{g - 1}{g} \end{aligned} \tag{3}$$

where

$$\kappa = \left(\rho\, c^2\right)^{-1}; \quad h = \frac{c_2}{c_1}; \quad g = \frac{\rho_2}{\rho_1} \tag{4}$$

This formulation is simplified to a line integral for underwater target that is axis symmetric at any point along the deformed axis. The line integral for finite-length cylinders is given by Refs. [14, 15]

$$f_{bs} = \int_{r_{pos}} \frac{k_1^2 a}{4k_2} \left(\gamma_k - \gamma_\rho\right) e^{2ik_2 r_{pos}} \frac{J_1(2k_2 a \ \cos\beta_{tilt})}{\cos\beta_{tilt}} |dr_{pos}| \tag{5}$$

where J_1 is Bessel function of the first kind, θ is incidence angle, k is incident wave number $= 2\pi/\lambda$, and λ is acoustic wave length. Target strength (TS) is the logarithmic of the backscattered signal

$$TS = 10\ \log \sigma_{bs} = 10\ \log |f_{bs}|^2 \tag{6}$$

where $\sigma_{bs} = |f_{bs}|^2$ is the backscattering cross section and f_{bs} is backscattering amplitude.

3. Results and discussions

Beam pattern of transducer in linear and decibel scales were shown in **Figure 2**. The main lobe has a higher power of about 40 dB from the first side lobes. This pattern is determined by acoustic frequency, size, shape, and phase of transducer. Maximum sensitivity of transducer along the main acoustic axis is 0 dB. Amplitude of side lobes is ranged from −80.0 to −40.0 dB. The maximum detection range of the echo sounder has been computed using signal to noise ratio, TS, frequency, electro acoustic efficiency, and acoustic power [16]. **Figure 3** shows that the detection range of echo sounder is about 220 m in depth and detectable breadth is 8 m from the acoustic axis. The noise resulted by research vessel is the largest because of the propeller noise. Signal to noise ratio (SNR) is the ratio of the echo power of the fish to the received noise power. Theoretical sphere target strength was numerically simulated for a 38.1-mm-diameter sphere of tungsten carbide. Theoretical and measurement of sphere ball target strength were shown in **Figure 4**. This figure explains that the measurement was suitable with theoretical value.

http://dx.doi.org/10.5772/intechopen.69156

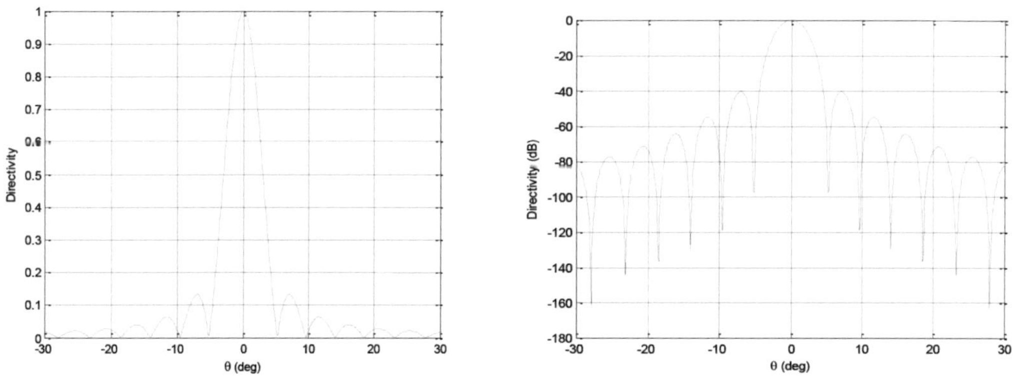

Figure 2. Beam pattern of transducer in linear (left) and decibel scale (right).

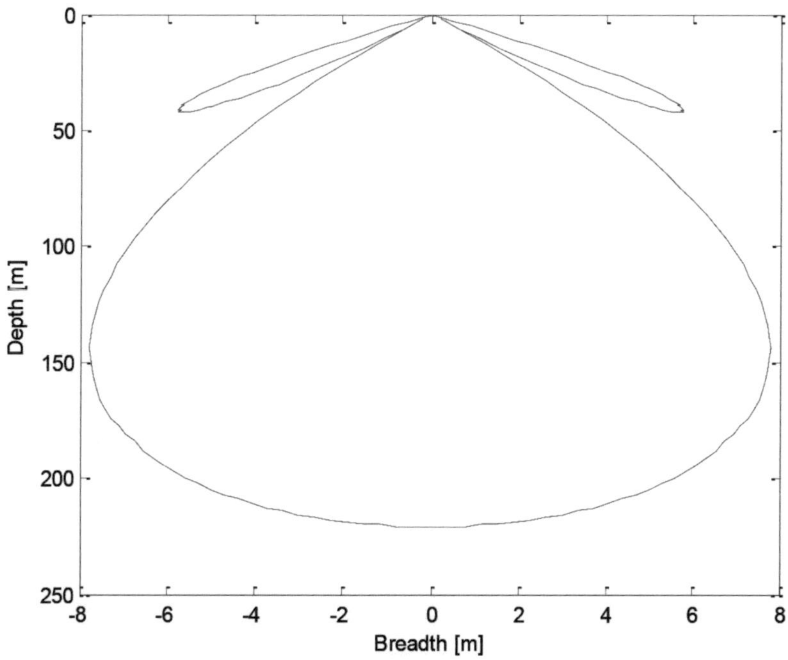

Figure 3. Detection range and detectable breadth of transducer.

Transmission loss measurement was shown in **Figure 5**. Increasing sound propagation range was followed by increasing transmission loss. The acoustic intensity/energy loss is due to spherical or geometrical spreading and attenuation. Acoustic ray propagation and its sound intensity level in several transducer depths were shown in **Figures 6** and **7**. The refraction of sound was caused by temperature gradients in the water, reflection from sea surface, sea bottom, and position of the target. Small changes in the temperature have significant influence on sound propagation. Acoustic detection of fish and seabed in the raw signal echogram and after filtering were shown in **Figures 8** and **9**, respectively. Target strength of fish ranged between −53.0 and −32.9 dB was shown in **Figures 10** and **11**, and volume backscattering signal was shown in **Figure 12**.

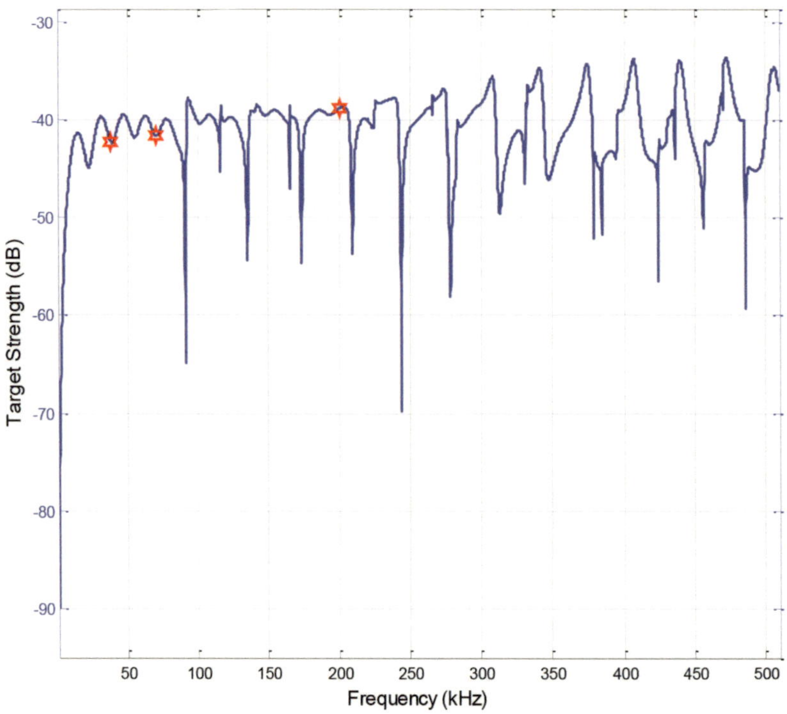

Figure 4. Measurement (*) and theoretical target strength (−).

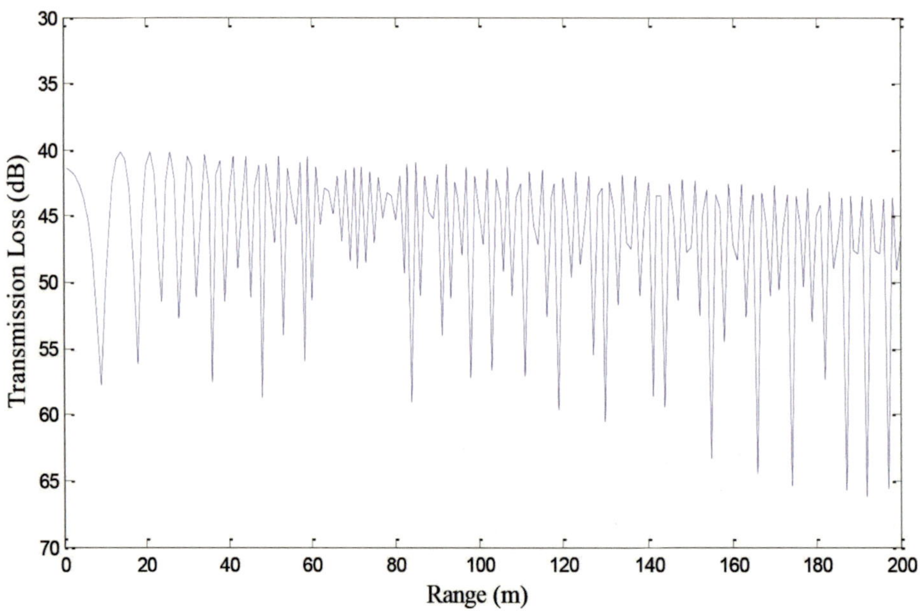

Figure 5. Sound transmission loss.

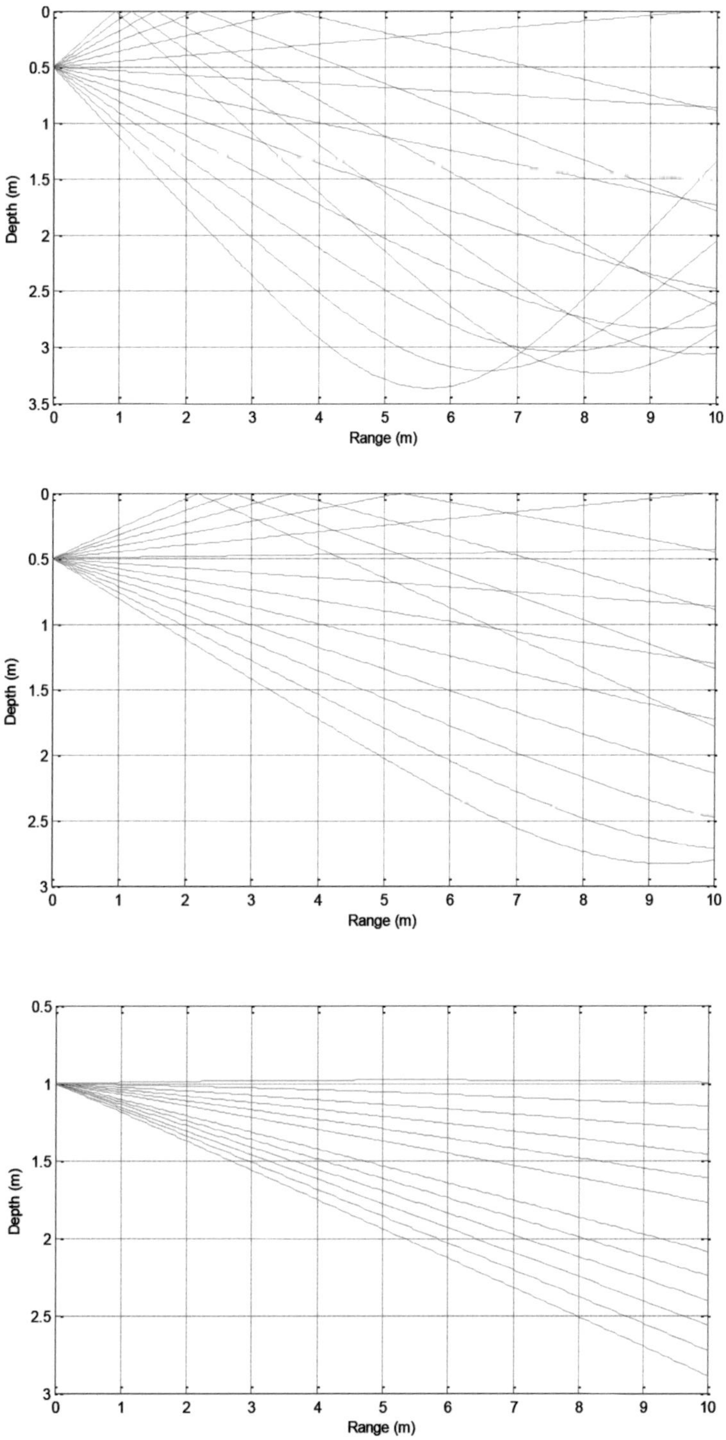

Figure 6. Acoustic ray propagation.

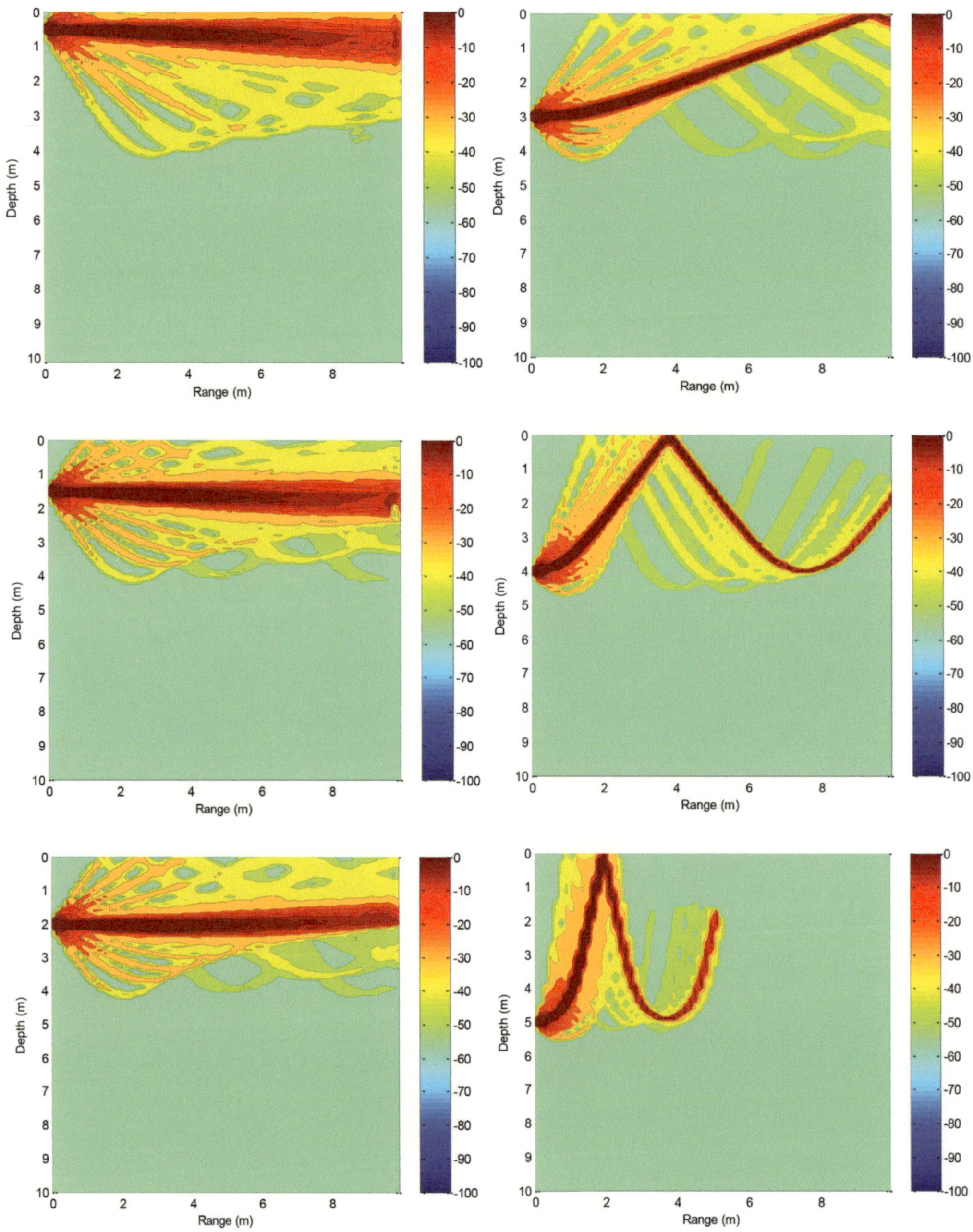

Figure 7. Sound intensity level for transmitter depth of 0.5, 1.5, 2, 3.0, 4.0, and 5.0 m.

Measurement of target strength (TS) in laboratory was conducted using 10 dead fish. The TS value for fish was determined by the tilt angle and acoustic frequency. The values of TS_{max} and TS_{avg} as functions of linear value of fish length are plotted in **Figure 13**. The values of TS_{max} and the TS_{avg} at 50 kHz were higher than those at 200 kHz. Positive

http://dx.doi.org/10.5772/intechopen.69156

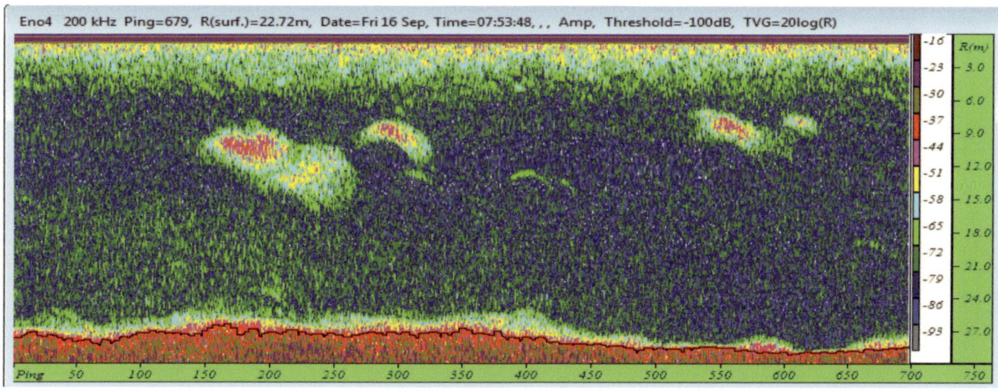

Figure 8. Raw data echogram.

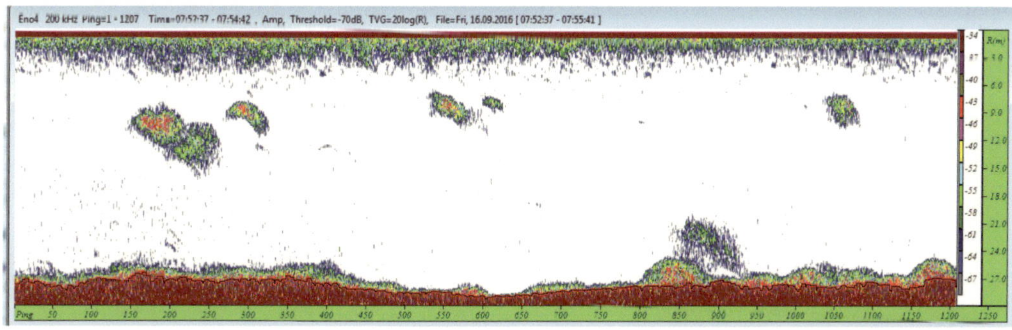

Figure 9. Echogram filtered.

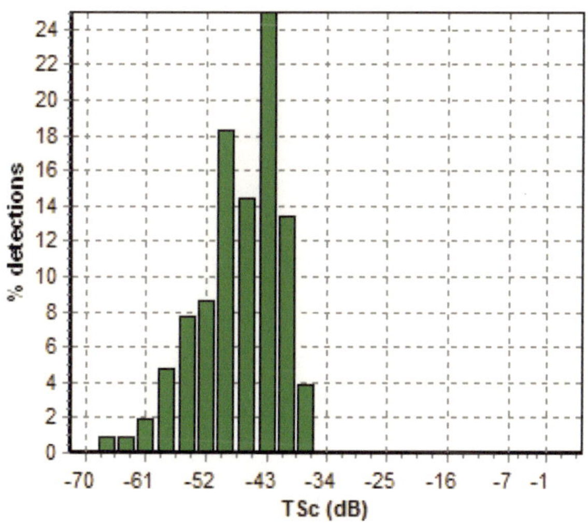

Figure 10. Target strength histogram.

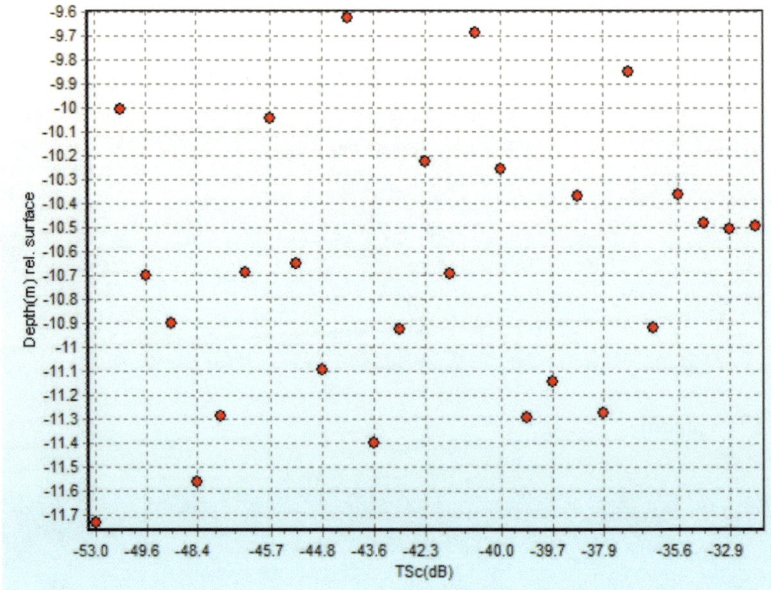

Figure 11. Target strength versus depth.

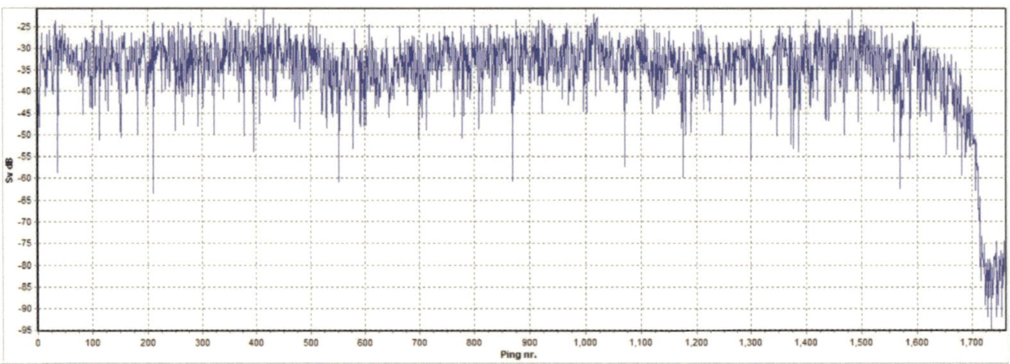

Figure 12. Volume backscattering (SV) signal.

correlation was found between TS values and fish length at both 50 and 200 kHz. The best fit regression lines of TS_{ave} are $TS_{ave} = 19.81 \log (FL) - 98.2$, $r = 0.96$ (**Figure 13**; left side) and $TS_{ave} = 19.56 \log (FL) - 96.47$, $r = 0.96$ (**Figure 13**; right side). A small discrepancy was found in TS_{max} and TS_{ave}. The slope of TS_{max} was close to 20, suggesting that the acoustic backscattering was proportional to the square of fish or body length. For TS quantification, acoustic threshold was applied (**Figure 14**), and application of single echo detector was shown in **Figure 15**.

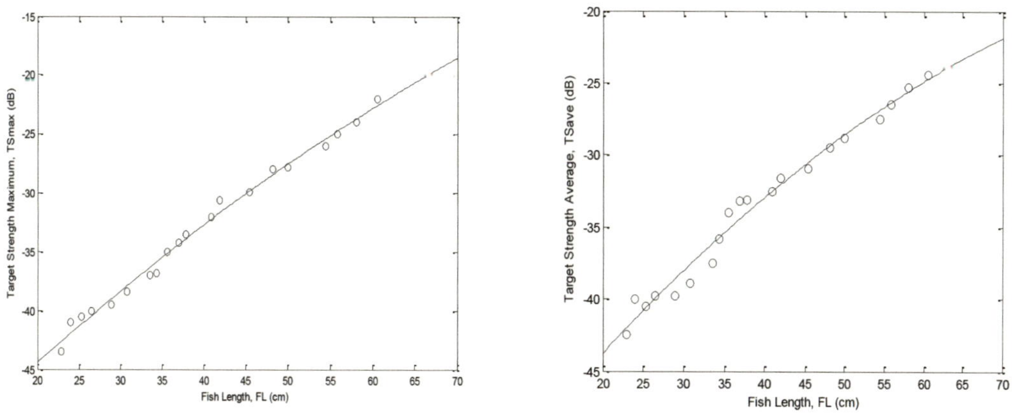

Figure 13. Relationship between TS and fish length (FL).

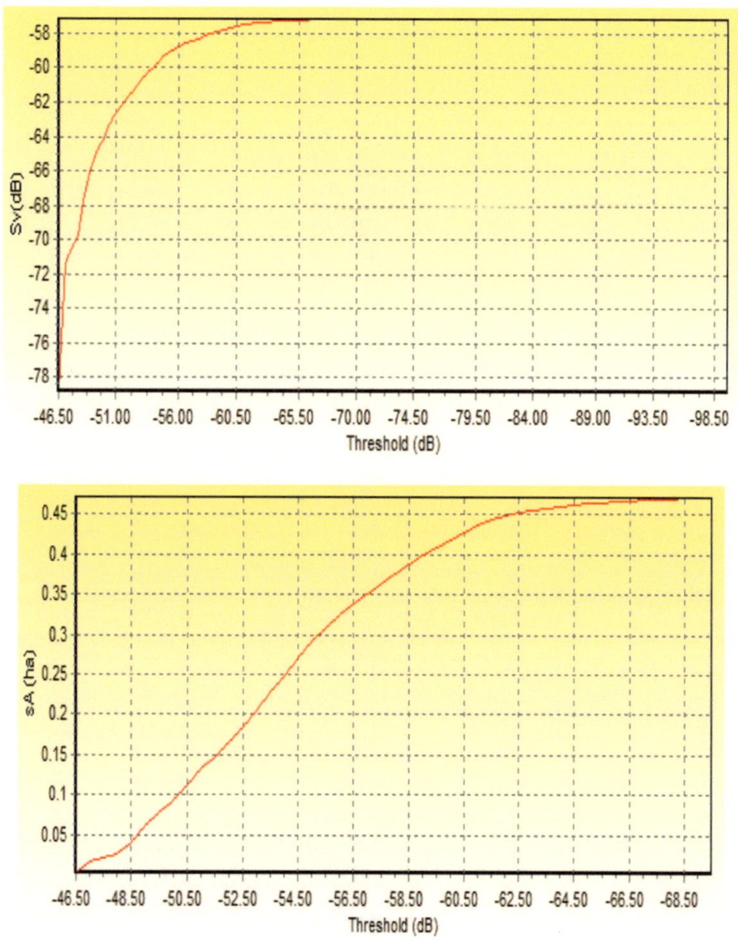

Figure 14. Threshold application for SV and SA modes.

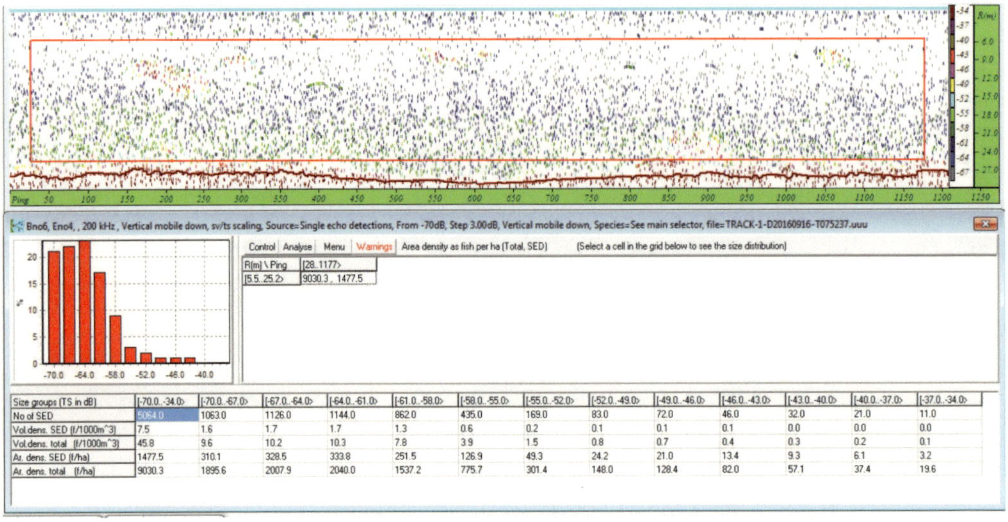

Figure 15. Single echo detector for TS detection.

Typical examples of TS as a function of incidence angle at frequencies 50 and 200 kHz are shown in **Figure 16**. The variations of TS value with incidence angle are displayed at 0° (main lobe) at both frequencies. The side lobes are displayed at a small discrepancy at two frequencies. The peaks were sharp, suggesting that slight changes in the incidence angles of fish have a major effect on the TS value.

Target strength of fish is important for fish stock estimation. The measurement of fish density uses TS as a scaling factor and instrument parameters. In fact, individual TS depends upon physical and biological factors such as tilt angle, length, acoustic frequency, physiology, and morphology [17].

Acoustic backscattering using the DWBA model requires accurate values of sound speed and density of fish. This is caused by a weakly scattering organism whose material properties vary from surrounding water. Acoustic scattering predictions with the tilt angle are measured for fish of angle increment from 0 to 360°. The comparison between DWBA model and measurement was agreed upon on the main lobe, but in the side lobe, there is some discrepancy. It was found the acoustic backscattering is strongly dependent on incidence angle and frequency. This result is suitable for the previous research using DWBA for zooplankton and squid applications [18, 19]. Target strength for several fish were shown to increase significantly from 0° to 90° and from 180° to 270° for all frequencies. In the future, the phase parameter of DWBA should be included in TS computation. This is the first research to measure the incidence angle of Indonesian fish in an experimental water tank and ocean field to apply a theoretical target scattering model using DWBA. We confirm that application of single-beam echo sounder is possible for accurate TS measurement.

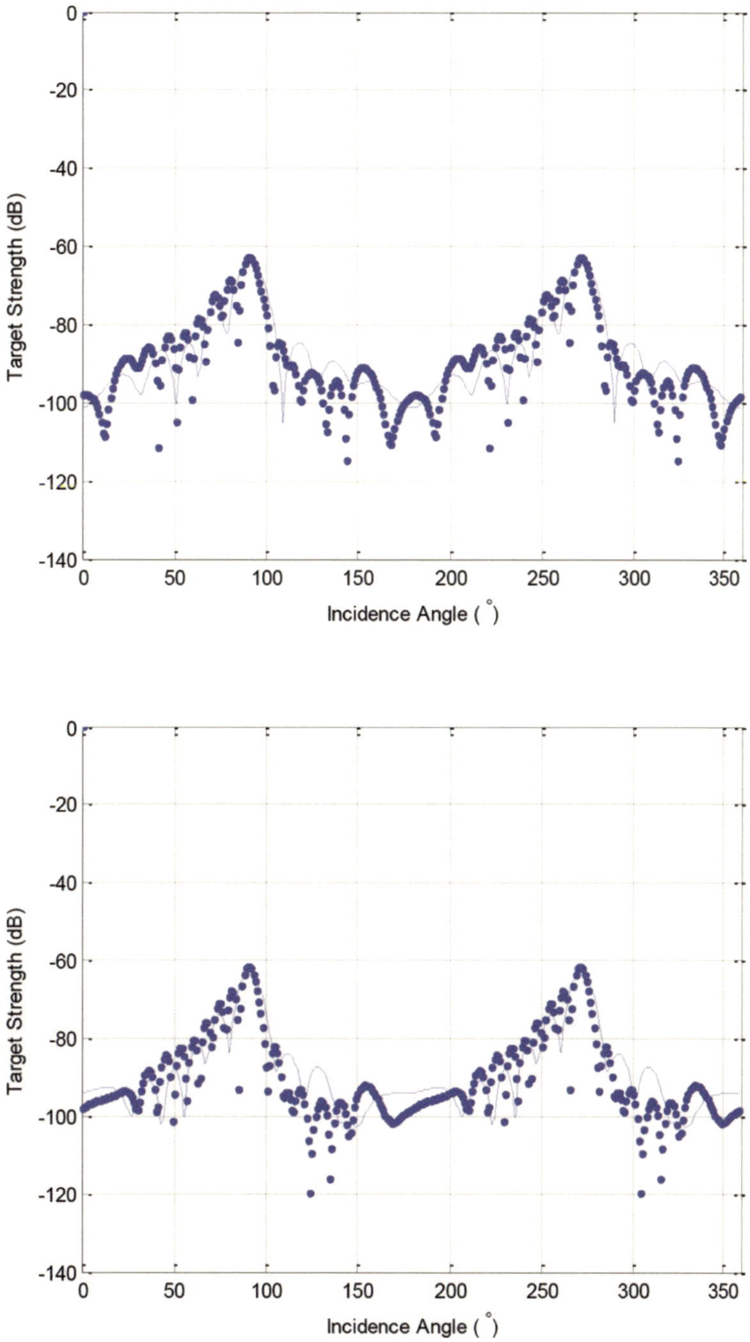

Figure 16. DWBA numerical model (-) and measurement (•) of TS values as a function of tilt angle at 50 (upper) and 200 kHz (lower).

4. Conclusion

The results indicated that TS of fish was determined by incidence angle of acoustic wave, fish length, and frequency of sonar instrument. TS will increase with the length of the animal. TS information are useful for quantifying fish stock in the field using quantitative echo sounder. The validation of DWBA model to measure target strength is confirmed with the laboratory experiment using single-beam echo sounder.

Acknowledgements

We acknowledge the Ministry of Research, Technology, and Higher Education Indonesia for financial support of this research.

Author details

Henry M. Manik[1]*, Dony Apdillah[2,3], Angga Dwinovantyo[4] and Steven Solikin[4]

*Address all correspondence to: henrymanik@ipb.ac.id

1 Department of Marine Science and Technology, Faculty of Fisheries and Marine Sciences, Bogor Agricultural University (IPB), Indonesia

2 Graduate Student of Marine Technology Bogor Agricultural University (IPB), Indonesia

3 University of Maritime Raja Ali Haji (UMRAH), Indonesia

4 Graduate Student of Marine Technology PMDSU Program Bogor Agricultural University (IPB), Indonesia

References

[1] Manik HM. In: Kolev N, editor. Underwater Acoustic Detection and Signal Processing Near the Seabed, Sonar Systems. InTech; 2011. DOI: 10.5772/17499. Available from: http://www.intechopen.com/books/sonar-systems/underwater-acoustic-detection-and-signal-processing-near-the-seabed

[2] MacLennan DN, Simmonds EJ. Fisheries Acoustics. London: Chapman & Hall; 1992

[3] Foote KG. Importance of the swimbladder in acoustic scattering by fish: A comparison of gadoid and mackerel target strengths. Journal of the Acoustical Society of America. 1980;**67**(6):2084–2089

[4] Ona E. Physiological factors causing natural variations in acoustic target strength of fish. Journal of the Marine Biological Association of the United Kingdom 1990;**70**:107–127

[5] Ona E, Mitson RB. Acoustic sampling and signal processing near the seabed: The dead zone revisited. ICES Journal of Marine Science. 1996;**53**:677–690

[6] Manik HM. Measurement of acoustic reflection of tuna fish using echosounder instrument. ILMU KELAUTAN: Indonesian Journal of Marine Sciences. 2009;**14**(2):84–88

[7] Manik HM. Measurement and numerical model of fish target strength for quantitative echo sounder. AACL Bioflux. 2015;**8**(5):699–707

[8] Manik HM. Quantifying fish backscattering using SONAR instrument and Kirchhoff ray mode (KRM) model. Journal of Physics: Conference Series. 2016;**739**:012055. DOI: 10.1088/1742-6596/739/1/012055

[9] Stanton TK, Clay CS, Chu D. Ray representation of sound scattering by weakly scattering deformed fluid cylinders: simple physics and applications to zooplankton. Journal of the Acoustical Society of America. 1993;**94**:3454–3462

[10] Manik. Acoustic observation of zooplankton using high frequency sonar. ILMU KELAUTAN: Indonesian Journal of Marine Sciences. 2015;**20**(2):61–72

[11] Balk H, Lindem T. Sonar 4 and Sonar 5 Post Processing Systems, Operator Manual Version 602. Oslo: Lindem Data Acquisition; 2014

[12] Chu D, Foote KG, Stanton TK. Further analysis of target strength measurements of Antarctic krill at 38 and 120 kHz: Comparison with deformed cylinder model and inference of orientation distribution. Journal of the Acoustical Society of America. 1993; **93**:2985–2988

[13] Morse PM, Ingard KU. Theoretical Acoustics, Chapter 8. Princeton: Princeton University Press; 1968

[14] Stanton TK, Chu D, Wiebe PH, Martin L, Eastwood RL. Sound scattering by several zooplankton groups I: Experimental determination of dominant scattering mechanisms. Journal of the Acoustical Society of America. 1998;**103**(1):225–235

[15] Stanton TK, Chu D. Review and recommendations for the modeling of acoustic scattering by fluid-like elongated zooplankton: Euphausiids and copepods. ICES Journal of Marine Science. 2000;**57**:793–807

[16] Furusawa M, Asami T, Hamada E. Detection range of echosounder. In: The 3rd JSPS International Seminar. Sustained Fishing Technology in Asia towards the 21st Century; 2000. pp. 207–213

[17] Manik HM, Nurkomala I. Measurement of target strength and fish stock in Pari Islands seawaters using single echo detector method. Marine Fisheries. 2016;**7**(1):69–81

[18] McGehee DE, O'Driscoll RL, Traykovski LVM. Effects of orientation on acoustic scattering from Antarctic krill at 120 kHz. Deep-Sea Research Part II. 1998;**45**(7):1273–1294

[19] Jones BA, Lavery AC, Stanton TK. Use of the distorted wave born approximation to predict scattering by inhomogeneous objects: Application to squid. Journal of the Acoustical Society of America. 2009;**125**(1):73–88